WOODFUEL MARKETS IN DEVELOPING COUNTRIES

Woodfuel Markets in Developing Countries
A case study of Tanzania

JILL BOBERG

LONDON AND NEW YORK

First published 2000 by Ashgate Publishing

Reissued 2018 by Routledge
2 Park Square, Milton Park, Abingdon, Oxon OX14 4RN
711 Third Avenue, New York, NY 10017, USA

Routledge is an imprint of the Taylor & Francis Group, an informa business

Copyright © Jill Boberg 2000

All rights reserved. No part of this book may be reprinted or reproduced or utilised in any form or by any electronic, mechanical, or other means, now known or hereafter invented, including photocopying and recording, or in any information storage or retrieval system, without permission in writing from the publishers.

Notice:
Product or corporate names may be trademarks or registered trademarks, and are used only for identification and explanation without intent to infringe.

Publisher's Note
The publisher has gone to great lengths to ensure the quality of this reprint but points out that some imperfections in the original copies may be apparent.

Disclaimer
The publisher has made every effort to trace copyright holders and welcomes correspondence from those they have been unable to contact.

A Library of Congress record exists under LC control number: 99073643

ISBN 13: 978-1-138-71368-0 (hbk)
ISBN 13: 978-1-138-71367-3 (pbk)
ISBN 13: 978-1-315-19832-3 (ebk)

Contents

List of Figures vi
List of Tables vii
Acknowledgments ix

1 Introduction: Woodfuel, People and Environment 1

2 Woodfuel Markets 22

3 Data Collection 42

4 Structure of the Woodfuel Market 53

5 Conduct of Market Participants 88

6 Market Performance – Productive Efficiency 97

7 Performance – Pricing Efficiency 108

8 Policy Implications and Conclusions 136

Bibliography 167

Appendix 174

List of Figures

Figure 3.1	Tanzania	43
Figure 3.2	Woodfuel supply path – thicker lines indicate more commonly used connections	45
Figure 4.1	Approximate woodfuel supply areas: Dar es Salaam, Mbeya and Shinyanga	56
Figure 4.2	Distance travelled by fuel	59
Figure 4.3	Transport modes for firewood	61
Figure 4.4	Transport modes for charcoal wholesalers	62
Figure 4.5	Transport modes for charcoal retailers	62
Figure 4.6	Supply of woodfuels – number of nodes in supply system for each city	64
Figure 4.7	Primary occupation and longevity of fuelwood producers	70
Figure 7.1	Cost structure for charcoal	111
Figure 7.2	Cost structure for firewood	112
Figure 7.3	Relationship between kiln price and distance to town and road condition	117
Figure 7.4	Real charcoal prices and trends	128
Figure 7.5	Real kerosene price	132
Figure 7.6	Electricity prices in Dar es Salaam	133

List of Tables

Table 3.1	Survey sample size	47
Table 4.1	Household woodfuel use	54
Table 4.2	Percentage of fuel by weight obtained by households from each source	55
Table 4.3	Fuel supply areas – percent of supply reaching end users	57
Table 4.4	Distances travelled by fuels	59
Table 4.5	Participant profiles	66
Table 4.6	Employment history, reasons for being in woodfuel business	67
Table 4.7	Primary income source of woodfuel producers	68
Table 4.8	Differences in years in business by fuel supply area	71
Table 4.9	Kilograms of fuel bought and sold yearly at each stage of the supply system	72
Table 4.10	Distribution of sellers by gross annual turnover	74
Table 4.11	Size distribution of charcoal sellers (by number of bags sold)	75
Table 4.12	Ratio of value added to total sales	78
Table 4.13	End user reasons for choosing a particular woodfuel seller	80
Table 4.14	Percent of participants who always sell to the same buyer or buy from the same seller	81
Table 4.15	Access to credit for participants who sell to more than one buyer	83
Table 4.16	Seller longevity	85
Table 5.1	Primary factor determining selling price of woodfuel sellers	89
Table 5.2	Sales promotion by sellers	92
Table 5.3	Extension of credit by sellers	94
Table 6.1	Efficiency factors for charcoal producers	99
Table 6.2	Capital turnover and days between supply purchases – charcoal traders	101
Table 6.3	Labour force for sellers	102

Table 6.4	Scale efficiency in charcoal sellers	103
Table 7.1	Fuel cost structure	109
Table 7.2	Return to labour	113
Table 7.3	Fuel price build-up – Dar es Salaam	114
Table 7.4	Fuel price build-up – Mbeya	115
Table 7.5	Fuel price build-up – Shinyanga	115
Table 7.6	Relationship between road distance and producer price	118
Table 7.7	Fuel pricing in Dar es Salaam districts	122
Table 7.8	Seasonal change in woodfuel price 1989–90	124
Table 7.9	Index of quarterly retail prices of woodfuel 1977–90	126
Table 7.10	Factors influencing charcoal prices	127
Table 7.11	Percent of participants agreeing with supply scarcity questions	130
Table 7.12	Annual changes in woodfuel prices 1989–90	131
Table 8.1	Differences between charcoal markets in three cities	144
Table 8.2	Differences between firewood markets in three cities	146
Table 8.3	Kerosene substitute price for woodfuel	159
Table 8.4	Economic costs of woodfuel	160

Acknowledgments

The study on which this work is based was part of a larger project undertaken in Tanzania in 1990-1991. The Stockholm Environment Institute and SIDA generously funded the Tanzania Urban Energy Project. Thanks to SEI's Lars Kristofferson and Sven Svenson for their help and guidance, and to Gordon McGranahan for his advice and ideas, particularly at the early stages of the formulation of the surveys and survey methodology.

The energetic director and author of the Tanzania Urban Energy Project, and my advisor, supporter, and intellectual guide was Richard Hosier. My grateful thanks go out to him for his generous support over the years, and his inspired captainship of the project through the usual potholes of international work (not to mention the roads of Dar es Salaam!). I also thank him for his patience with me and his ideas and editing which contributed so much to the work. I also thank Amy Hosier for editing help, and for laughs and commiseration during and after our stay in Tanzania. On On!

Many thanks also go out to the other leaders and conceivers of the Tanzania Urban Energy Project, Dr Mark Mwandosya and Dr Matthew Luhanga for the time they spent and support they gave to the project despite their many other duties.

Particular thanks go out to my friends and colleagues, Wilfred Kipondya and Theofilo Bwakea. They welcomed me to the country, were generous with their friendship as well as their hard work, and were instrumental in the success of my study and the project as a whole. Many other Tanzanian colleagues, too numerous to name, were a great help to my work. These include in particular the wonderful group of enumerators who accompanied me to the rural and urban reaches of Tanzania. They kept us laughing constantly, introduced me to the diverse cultures of the areas we were visiting (always looking for the best local brew and the cleanest guest house!), and were uncomplaining as I pushed them to meet deadlines and budget constraints. Thanks, guys!

1 Introduction: Woodfuel, People and Environment

Woodfuel, People and the Environment

There has been much written about the environmental problems which face developing countries. In a world newly awakened to the potential problems threatened by pollution, ozone depletion, and deforestation, developing countries are being drawn increasingly into the limelight. In Africa, the focus is on deforestation. The effects of the removal of vegetative cover from African lands vary from soil loss and lower agricultural productivity to watershed destruction and global climate effects. The destructive exploitation of forests in Africa takes a toll economically, socially and environmentally. At the local level, these effects of forest destruction translate into lowered incomes in the agricultural sector and a reduced standard of living. When the environmental effects of deforestation are combined with the acute economic problems suffered by most African countries, sustained poverty and human suffering are the logical outcomes.

Despite this negative forecast, such dire results are not inevitable. The challenge of stemming the environmental degradation and the declining living standards of people in African countries is one that can be met with knowledge and action. One of the areas in which such knowledge and action is needed is energy, and more specifically, energy from woodfuels. This is the area on which this study is focused.

There are good reasons to concentrate on woodfuels when exploring deforestation in Africa. Along with the extension of agriculture, the logging of trees for timber, and the overgrazing of cattle, one of the oft-cited causes of tropical deforestation in Africa is the cutting of trees for woodfuel.[1] The general consensus as to the activity on which lies the greatest blame for deforestation has shifted from fuelwood cutting to the clearing of land for agricultural use in recent years. Nonetheless, woodfuel energy is inextricably tied to the problem of deforestation. It is at times both a contributor to and a casualty of increased deforestation. As the availability of wood lessens, the cost of obtaining woodfuel increases, adding hardship to the lives of great numbers of people. In rural areas, the time spent collecting fuel increases with increased deforestation. Likewise, in cities the price of woodfuels increases when the cost of their supply grows as deforestation requires suppliers to go further afield to obtain the woodfuel.

Woodfuels (firewood and charcoal) are the major source of energy for most Africans, since they often have insufficient resources for or little access to alternative fuels, in particular electricity and petroleum-based fuels. Woodfuels supply anywhere from 50 per cent to 95 per cent of the total energy consumed in sub-Saharan African countries, with its dominance particularly acute in the household sector. This dependence on woodfuel means that the potential effects of deforestation as it relates to the provision of woodfuel for energy can be enormous. As always, the effects will strike most deeply at the poor.

Increased urbanization has led to increased concern over the effects of and problems associated with urban fuelwood use. The largest concentration of energy users is in the urban areas. Urban users, while generally having a choice of fuels, nonetheless often prefer fuelwood even when prices of fuelwood are higher than those of alternative fuels. Security of supply, small, affordable quantities, and the absence of a large initial investment in a cooking stove are all qualities of woodfuel use which influence users' preference for those fuels.

The price at which woodfuel is sold impacts consumers by determining the level of energy use they can afford, and hence their standard of living. Prices also affect a consumer's choice of fuels. This choice in turn impacts the environment by affecting the amount of wood that is harvested. Because woodfuel prices are communicated to the urban users by way of the market, and because users make their choices of the form and quantity of their energy use in response to price signals, the effectiveness of the market as a medium for transmitting signals to end users is important. If prices do not reflect the true value of the wood resource, in ecological, social, and economic terms, the resource will not be used in a way that is optimal for the welfare of the country and will result in long term unsustainability.

One of the ways price distortions in urban markets are engendered is by the manner in which woodfuels are procured for urban use. Urban dealers often destroy woodlands by clear cutting wooded areas without concern for the sustainability of those areas. Because they usually pay little or nothing for the trees that they cut, they essentially mine the wood resource, and pay only the costs of extraction (Mercer and Soussan 1992). Those who eventually buy this woodfuel in the urban area are also paying only for the labour and capital involved in cutting, transporting, and selling the fuel. In the case of charcoal, the cost of transforming the wood into charcoal is also passed on to the urban consumers. However, the value of the wood resource to the physical and human environment of the rural areas is not included in the price, and this lack encourages deforestation and over use of the fuels, and undermines the sustainability of the system.

Other types of distortions in the woodfuel market are also possible, and common. There is particular concern over the effects of urban woodfuel use on deforestation around an urban area. This situation is often caused or exacerbated by the lack of well-defined and enforceable property rights in supply areas. The problems in land tenure can lead to unsustainable exploitation of wood resources, and to the distortion in prices discussed above.

Both of these distortions affect the price of a fuel, and are important to the long term sustainability of the system. But there is another level of price distortion that is also important to the final price of the fuel and all that it implies. If the price of fuel were freed of the distortion caused by ill-defined property rights and biomass mining, there would still be the potential for price distortions for the urban user. The wood must be cut and processed and distributed to the users. This distribution system can impose its own distortions on the market and the price of the fuel if it does not work well. A market which distorts signals of scarcity to consumers can lead to overuse or under use of woodfuels, and overuse or under use of other fuels. This can have lasting effects on both the sustainability of the lands that provide the fuel and the survival of those least able to afford energy and other basic needs. If a fuel is cheaper than it should be, more people will tend to use it, and those who use it will tend to use more of it. If the fuel is a woodfuel, more trees will be cut down than would otherwise be the case, contributing to the destruction of the wood resource. Conversely, if the price of a fuel is too high, fewer people will use the fuel, and those who do will use smaller quantities of the fuel. This will keep people from making full use of the available resource, and can lead to social hardships as well as, for woodfuels, discourage activities that would, for example, enhance the wood resource. In this way it is important that the market, and the distribution system which is part of it, are free of distortions and other problems.

Although there has been much interest in the woodfuel sector in recent years, and large amounts of aid money has gone into solving the problems of its supply and use, there has been little progress towards that goal. Often, it has been thought that deforestation in Africa could be slowed significantly or halted as a direct result of controlling the amount of woodfuel energy used and supplied. The problem of deforestation was therefore often looked upon as an energy problem. As a result, most money and work has focused on the replacement of forests thought to be decimated as a direct result of energy provision, and on reductions in the use of woodfuel. Forests have been limited in their use, and reforestation schemes and supply augmentation in the form of plantations and woodlots have been vigorously promoted. Efficient charcoal kilns and cook stoves have been introduced, and alternative fuels and electrification have been advocated. Although there is merit in such policies, few of the programs initiated have achieved the effects originally envisioned. The lack of results from these attempts, either in the cessation of the trend toward deforestation or in the reduction of dependence on woodfuel energy, suggests that a different tack must be taken.

A result of the experience gained from these efforts has been a change of focus. It is now understood that while energy provision is not the dominant cause of deforestation, it is nonetheless tied to the problem and is of great importance to the economy and standard of living of African countries. There is a resultant interest in a more in-depth understanding of woodfuel and its provision, and its place in the complex structure of basic needs of people and of the society as a whole. Woodfuels need to be considered in light of their relationship to social issues such as the economy, food provision, population growth, land use, politics, and other issues, as well as their established relationship to the environment and environmental sustainability. Fuelwood problems are the result of fundamental failures in both the physical and human landscapes, in areas such as rural land, labour and capital markets and urban energy markets as well as the potential for regrowth of trees and increased woodfuel supply. Fuelwood problems are tied to failures in efficient and sustainable land allocation, which in turn are related to lack of the appropriate conditions being established by the government (Mercer and Soussan 1992). While this complicated interrelationship means that problems become ever more complex and difficult to resolve, it also means that any actions taken as a consequence are more likely to be successful in making substantive progress towards the resolution of the problems.

It also means that policies concerned with the environment and wood energy must be rethought. If earlier policies have not been effective, there is little with which to replace them. The next round of research and aid projects must be an attempt to come up with policies that will take into account the new view of an integrated problem.

The Present Study

The purpose of this study is to aid in the process of formulating effective new policies governing woodfuel supply in the context of the effect of woodfuel use on the social and environmental landscape of African countries. It will approach the subject from the viewpoint of the desirability of sustainability – the sustainability of woodfuels as the major provider of energy to poor and rich households and commercial enterprises in Africa; the sustainability of the forests and other sources of supply for this fuel; and the sustainability and improvement of the living standards of the users of woodfuel.

In order to accomplish this goal, the research will take a case study of the woodfuel situation in Tanzania. It will focus on the urban aspects of woodfuel use, and within that, on the market for the fuel and its effects on price. Despite the importance of woodfuel to developing country energy economies, very little is known about the system that supplies woodfuel to the urban sector. The role of distributional efficiency in the structure and operation of the system has been little studied. This is despite the fact that the overall competitiveness of the system, and its ability to transfer messages about the value of the fuel through market prices is important not only to consumers but also to policy makers who must make informed policy decisions. One of the missing elements in previous research has been the data and analytical framework to do a thorough investigation of the network through which woodfuels are moved from harvester to end-user. This research will contribute both a solid set of data on Tanzania's woodfuel marketing system, and a systematic analysis of those data to determine how well that system functions, and the effect of any malfunctions on the people and environment of Tanzania. After analyzing the marketing network and determining its effects on price, the impact of this information on this type of market distortion will be examined in the light of other market distortions such as the 'free' cutting of wood for woodfuel. Then it will examine policy possibilities that can deal with the combined effect of these distortions.

The remainder of this chapter will describe in more detail the importance of woodfuel to Tanzania. First, the country's dependence on woodfuels will be highlighted. Next, the interaction of the use of woodfuels and deforestation will be discussed, pointing out particularly the situation as it stands in Tanzania. That will be followed by a more detailed look at issues particular to urban energy provision. Finally, the importance of markets to the provision of urban woodfuel, their function as price signalling mechanisms and their ultimate effect on the social and environmental atmosphere will be discussed.

Dependence on Woodfuel

The vast majority of people in many developing countries, including those of sub-Saharan Africa, rely exclusively or partially on biomass fuels (including firewood, charcoal, and crop or animal residues) to meet their energy needs. In fact, fuelwood supplies more than 75 per cent of the total energy used in many countries, including many if the poorer countries of Africa and Asia such as Nepal, Bangladesh, Ethiopia and Burkina Faso (Soussan 1988). Woodfuel accounts for 80 per cent of the total energy consumed in the countries of Southern Africa (Munslow et al. 1988). The majority of the energy consumed in these countries is used in the household sector, for cooking.

Tanzania also depends heavily upon biomass energy to fuel its subsistence energy needs. Fully 90 per cent of the energy consumed in Tanzania is consumed in the household sector (Hosier and Kipondya 1993). Even in urban households, which are more likely to use 'modern' fuels than their rural counterparts, 84 per cent use woodfuel. Overall, almost 90 per cent of energy consumed in Tanzania is from woodfuels (Victus 1993). In industry, wood provides 37 per cent of the energy consumed, and Tanzania's informal sector uses woodfuel almost exclusively as its energy source (Hosier 1993).

The dependence on woodfuels evident today in developing countries is likely to increase. Populations are growing, with the population of sub-Saharan Africa currently increasing at a rate of 3.1 per cent per annum, and Tanzania's population growing at the same rate (World Bank 1992). The poor have few alternative energy sources, particularly in the rural areas. The need for woodfuels and the environmental effects of the use of woodfuels and, equally, the effects of deforestation on access to woodfuels are likely to remain constant or to increase with population and will continue to be of concern in developing countries for the foreseeable future.

Woodfuel and the Environment

Deforestation and Woodfuel

Recent concern over deforestation in the world community is centred on the loss of large areas of tropical forest to development pressures. The areas devastated are generally large, uninhabited tracts of forest far from populated areas. Not only is the harvesting of trees for use as fuelwood not responsible for such wholesale land clearance, but the initial land clearance attracts migrants who continue to clear land for agricultural purposes, producing a surplus of woodfuel for which there is no demand (Mercer and Soussan 1992).

It is generally agreed that rural woodfuel use, culled mostly from trees on or near farms outside forest areas, is not responsible for large-scale deforestation. Local woodfuel use may contribute to the degradation of land resources in areas that are vulnerable environmentally and economically, but other factors are normally the driving force behind the original resource degradation. The interaction of competing land uses and the environment creates a situation that is unsustainable in the long run.

Land clearance due to growing populations and the commercialization of rural economies is the major culprit contributing to deforestation in rural areas. Farming, both subsistence and commercial, and animal grazing are the most frequent uses of the cleared land. Over clearing for these purposes puts pressures on wood resources used for other purposes such as fuel, fodder and building materials.

Urban fuelwood and charcoal markets can also contribute to such pressures. Although not the villain it once was believed to be, the harvesting of forests to provide woodfuel for urban areas has been cited as a contributing factor to deforestation in some regions.[2] However, the bulk of the tree cutting done to supply urban areas is done around cities and in forest areas close to roads and railways. Usually, trees are cut selectively, since particular species and sizes are preferred for use as firewood and in the production of charcoal. As a result, land containing trees cut specifically for the supply of woodfuel is often cut sequentially over a number of years, as trees initially considered too small to cut grow to adequate size, and new trees grow or coppice to replace those cut earlier. In selective harvesting, the understory of small trees and brush is left intact. This contrasts with the clear cutting that takes place when preparing land for agricultural use, and leaves the soil completely exposed to the elements.

Sometimes, the land that has been partially cleared by harvesting for woodfuel is further cleared for agricultural use. As well, some of the wood used to supply urban areas is cut originally as part of agricultural clearing. This pattern applies generally to many countries, and has been observed in areas of Tanzania.

There are significant environmental differences between the effects of selective harvesting and clear cutting. Land that is completely cleared will suffer greater erosion, causing sediment to invade streams. It will store less moisture in the soil, cause chemical changes in the soil, and will consequently result in slower regrowth of trees (Hosier 1993, Lewis and Berry 1988, Allen

1985). In areas with little regenerative potential (especially low rainfall areas), a series of clearings and poor after-harvest management may finally result in irreversible damage, and deforestation (Hosier 1993).

Deforestation and woodfuel in Tanzania As part of the project on which this research is based, past and present charcoal production sites in Tanzania were visited and assessed as to the environmental damage caused by the harvesting of wood for the production of charcoal. The results reported by Hosier (1993) showed a difference between sites that were clear-cut and those that were cut selectively. Erosion was minimal in sites that were selectively harvested and more substantial in clear-cut sites. Recovery from harvesting was vigorous in the areas studied, and many sites were used over and over again as trees regrew and were reharvested. This occurred even in areas that were cleared for agricultural use. After crops were grown, trees again took possession of the cleared area. The recovery times seemed to be longer than for selectively harvested areas. Overall, selective harvesting appears to be a fairly benign and sustainable method of harvesting trees for woodfuel.

However, woodfuel harvesting is not the sole use of most woodland areas. Human intervention, both by use and by management, can affect the sustainability of the woodlands in any area. If total use is intense, deforestation can occur, particularly in low productivity areas. In Tanzania, Shinyanga district is an example of a multi-use low potential woodland area that has been virtually deforested by overuse. Many years of clearing for intense cultivation and over-grazing have resulted in serious erosion and a reduced or arrested ability of the woodlands to recover (Hosier 1993). Human intervention has had a positive effect in Mbeya, where managed plantation woodlots within the city have supplied much of the population with fuel for many years. A third example is in the hinterlands of Dar es Salaam, the largest city in the country. There, high potential land (vegetation rich, adequate rainfall), selective harvesting and communal management have combined to provide what seems to be a fairly resilient woodlands system.

In general, a pattern emerges in determining areas that are particularly susceptible to deforestation. In Tanzania, provinces which had deforestation problems were likely to be one of three kinds: urban hinterlands; high potential, high rainfall areas (densely populated and farmed); and low potential, low rainfall areas (low population, vegetation poor areas) (Hosier et al. 1990). However, positive human intervention can reduce damage to such areas. Although woodfuel harvesting is not the sole, or usually even the primary cause of deforestation, it is part of the rubric of human and environmental interaction that determines the long-term sustainability of forests and woodlands. Consequently, better management and understanding of woodfuel harvesting can only improve the chances for sustainability.

Effects of Deforestation

There are many effects of a decline in wood supplies. Environmental effects include loss of land productivity, watershed destruction, and silting of streams and dams as a result of erosion following deforestation. Additionally, the myriad uses of the forest are affected. The wooded areas are used for fodder, for poles for building, medicines, food, home for game, shade, and other uses not listed here, as well as for fuel. All of these are made difficult or impossible with a reduction in forest cover.

But the effects of deforestation go beyond the direct environmental consequences. Social consequences in the rural areas are also heavy. They include the necessity of devoting more labour time to collecting fuelwood, the added burden of which generally falls on women (Tinker 1987). Labour spent collecting firewood during agricultural seasons is often labour taken away from agricultural production (Munslow et al. 1988).

Another social effect of decreasing woodfuel availability is reduced energy use, which eventually leads to reductions in cooking, heating and boiling water for drinking, with potential health effects (Cecelski 1985, Munslow et al. 1988). Overall standards of living can decrease, especially as rural people dependent on collecting their own fuel are forced to spend cash on fuel purchases.

The effect of deforestation on prices In the urban areas, the most obvious result of the depletion of woodfuel resources is to raise the price of woodfuels and other forest products to consumers in the urban area. These prices are expected to rise as the transport distances and time taken to collect and deliver the fuel increase. As in the rural areas, price rises can lead to reduced fuel use, including cutbacks in meals cooked, drinking water boiled, and heating used. Since woodfuel is a cash commodity in the urban areas, all increases in price will take away from the standard of living of cash-poor urban consumers. Nevertheless, when the cessation of woodland depletion is the sole goal, an increase in urban consumer prices for wood products is not discouraged. Higher prices are expected to encourage more efficient use and encourage consumers to switch to other fuels, thereby decreasing pressure on wood resources. In fact, these prices are felt to be lower than they should be in terms of the environment, because they do not reflect costs, or externalities, imposed on the environment by the removal of trees from woodlots. This conflict between environmental and social concerns is an important one, and one which will be returned to in this discussion.

Environmental externalities Deforestation has many costs, many of which have been mentioned in the previous discussion. Some of those costs are directly environmental, such as soil depletion and watershed destruction, while others are social, such as increased transport distances for woodfuel. All have an economic cost in the short or long term.

These costs to third parties, or externalities, are costs which are not considered when a tree is cut or sold. For that reason, there is a failure in the market, and the tree is underpriced or undervalued by society. The amount by which the tree is underpriced can be estimated by assessing the indirect costs of cutting the tree. Some of these costs include: the cost of lost agricultural productivity; the cost of a loss of hydroelectric potential; the cost of lost water supply; the cost of a loss of forest products; and the permanent release of carbon into the atmosphere, affecting the global climate, where trees are replanted or allowed to regrow. The assessment of these social costs of the cutting of trees for firewood is not an easy task, but ignoring it results eventually in overuse of the woodfuel resource.

Stumpage fees Increased stumpage fees have often been suggested as an economic solution to the problem of the environmental and social externalities of tree harvesting. Stumpage fees are the amount of money collected for cutting a standing tree. They compensate the owner of the tree (often the government) for the right to harvest the tree, and, ideally, reflect its value to society (Openshaw and Feinstein 1989). Often, the stumpage fees in developing countries are extremely low, and at times nonexistent. Because of the nature of the wood resource, stumpage fees are by definition difficult to collect, and therefore even those that are officially levied often go uncollected. Additionally, there is a tradition in rural areas of wood from communal or government managed lands being free.

The optimal stumpage fee is difficult to determine. Although several methods have been suggested, there is no consensus as to which method is the most accurate. Additionally, data is often not available to make a complete assessment of the estimate of the optimal fee level. The various methods will be discussed later in this study.

The levying and determination of economically appropriate stumpage fees has received great attention as a solution for the overuse of woodlands by much of the international donor community, most noticeably the World Bank. However, even within the development community, there has been no consensus as to the appropriateness of stumpage fees, the methods by which to determine them or the breaking down of the considerable social and practical barriers that exist to their implementation. They have been one of many policy responses to the problem of mitigation of deforestation, and sustenance of energy supplies in developing countries.

Previous Policy Responses

The connection of woodfuel use to the reduction of forest cover, and the connection between deforestation and scarcity and dire environmental and social consequences has produced a variety of responses in the past twenty years. Most policy responses have approached the woodfuel problem as a simple energy supply and demand problem, an approach which has since been questioned (Foley 1988, Leach 1988, Leach and Mearns 1988, McGranahan 1986, Munslow et al. 1988, O'Keefe and Munslow 1988, Teplitz-Sembitzky and Schramm 1989). From this perspective, a projected demand for woodfuel greater than projected supplies produces a 'gap' requiring policies that decrease demand and increasing supply.

Neither supply-enhancement policies (tree plantations, social forestry, forest management, pricing) nor demand-reduction policies (improved stoves, improved charcoal kilns, alternative fuel subsidies, electrification) have been particularly successful. Although such policies can be helpful, and there is a place for some of them in areas experiencing scarcity, they have not solved the problems of fuelwood scarcity and woodland depletion they were originally intended to solve. This failure indicates that it may not be the solutions that have been misconceived, but the problems themselves. It is beginning to be widely understood that fuelwood problems are

> manifestations of more fundamental failures in rural land, labour and capital markets, urban energy markets, and failures of governments (local and national) to establish the conditions that would allow efficient and sustainable allocation of land and resources between forest and cropland and wood and food production (Mercer and Soussan 1992, p. 185).

In other words, what is needed is an integrated approach which looks at woodfuel in the context of the human and natural resource system, encompassing environmental degradation, institutional reform and social equity.

Access to Fuelwood Resources

The failure of governments to facilitate and provide conditions that allow efficient allocation of land between crops and forests is one of the ways in which unequal access to fuelwood resources is encouraged. This results in confusion of land tenure and ownership rights over biomass resources. Other factors that affect access to fuelwood include the location of the resource and management of the resource.

Rights to Land and Trees

Rights to land, and therefore to the trees that are growing there, are an important social aspect of the woodfuel problem. The presence of such rights, or tenure, if they are well defined, exclusive, secure, enforceable and transferable, are an important incentive for the sustainable maintenance of the land and its resources (Mercer and Soussan 1992). Not surprisingly, land over which a person has no control lends no particular incentive to sustain it. Even joint rights can be enough to insure sustainable land management. Confusion or insecurity as to land tenure is an institutional failure that can be devastating to the environment and to tree stocks (McGrath 1989, Mercer and Soussan 1992).

Land tenure can be categorized in one of three general categories: locally-owned private farmland; state or commercially owned land controlled from somewhere outside the local area; and communal land, owned by the state but open to use by the community or population at large.

Much of the fuelwood resource in rural areas is obtained from private land, from farms and, sometimes, from personal plantations. In much of Africa, as in Tanzania, no land can legally be owned by anyone but the state, but *de facto* ownership is generally conferred on anyone who farms or builds on a piece of land. In essence, then, much of the established farmland fits into the first category of land tenure.

The second land tenure category includes state forests and reserves and large commercial plantations. Access to and fuel collection from such lands is prohibited. Special permits are needed to collect fuel in these areas. In the case of woodfuels, commercial enterprises or other urban dwellers often hold the permits. The large commercial interest or the government, neither of whom is usually accessible to people living in the areas contiguous with the forest, holds control over these lands. Although the interests holding the land attempt to police it, the land is often trespassed on.

The final type of land tenure is communal land. This land is held communally, and is open for use by anyone. This is where a part of the fuelwood for use in rural areas is obtained, to supplement that gathered from farms. When such lands are under the control of local groups, the communal use of these lands is often sustainable. As local governance is eroded due to changing social or economic pressures, however, the land is more likely to be over-exploited or converted into farm or rangeland indiscriminately (Mercer and Soussan 1992, Clarke and Shrestha 1989).

This mix of land tenure types results in unequal access to wood resources. The tenure status of a particular piece of land may be uncertain, or the owner of the land may be remote or relatively abstract (such as the national government), and enforcement of rules governing remotely owned land is difficult. In this case, land use may be neither sustainable nor in the best interests of either the local community of the country as a whole. Any improvements in this area must be made through governmental reform, by better defining and redefining land tenure categories, and perhaps by redistributing the access rights to trees in some areas.

Urban Issues

The evolution of fuelwood markets in developing countries is a direct result of the urbanization that has taken place at a rapid rate over the past twenty-five years. In sub-Saharan Africa, the annual urban growth rate averaged 5.9 per cent from 1980-1990, as compared with 5.8 per cent from 1965-1980 (world averages over the same periods were 4.5 per cent and 2.6 per cent) (World Bank 1992). Tanzania's urbanization rate was almost twice that of Africa in general, at 10.8 per cent for the 1980-87 period, and 11.3 per cent for the 1965-80 period (World Bank 1992). Most of the growth comes from migration to urban areas from the rural areas. These migrants tend to be poor and to bring with them the tradition of woodfuel use from the villages.

Urban-rural Equity

Cities are able, by their size and political and economic clout, to use and abuse the resources of the rural areas. The reach of the cities can extend many kilometres into the countryside, into the most rural of areas. They can demand resources such as fuel and agricultural products and pay little for them, leaving people in the rural areas to absorb any environmental or other costs that are not reflected in the price of the product. On the other hand, the cities are filled with people who are too poor to be able to endure any increases in the cost of their basic needs, and any rise in prices can bring about great hardship for the urban poor and analogous political hardship for the government. In order to keep prices of woodfuel down, traders attempt to procure their product at the lowest possible cost, encouraging them to bend and break rules governing forest management. They often clear wooded areas and make no efforts to conserve or replace the resource, at times leaving little of the resource for the rural people who traditionally relied upon it. They essentially mine the resource, while paying only for the costs of extraction (Mercer and Soussan 1992).

Hence, a conflict has arisen between the rural areas and the urban areas. The conflict is not straightforward, because there is another side to the relationship between rural and urban. Many farmers and other rural dwellers earn their livelihood by supplying the cities with agricultural products and woodfuel and other tree products. Thus rural people who are within the growing range of economic influence of urban areas are dependent on the cities for income, but are paying for that dependence in many cases by the destruction of forests and other resources on which they depend to survive. Any solution to the problems posed by woodfuel use must thus take into account the economic and environmental needs of rural people, as well as the needs of the urban poor. For any solution, the equity balance between rural and urban dwellers, as well as between low and high income residents of the urban areas must be considered.

Fuel Choice

An outstanding feature of the urban energy scene is the concept of a fuel transition. As urbanization increases, and as income rises, people begin to switch to modern fuels from woodfuels (Leach 1992). The switch is often not complete, however – many people use several different fuels for different household applications. In Tanzania, this is the case – even in households that have electric stoves, some of the cooking is done using charcoal or kerosene (Hosier and Kipondya 1993). Although price is the most important factor regulating amount and type of energy used, households often choose to use woodfuels even if they are more expensive on an energy basis than modern fuels. The reasons for this are numerous. One of the most important reasons is security of supply. Woodfuel tends to be available at all times, while modern fuels are notorious for having supply interruptions. Also, woodfuel can be purchased in one-day or even one-meal quantities, and requires no expensive stove to burn. In addition, people, particularly new arrivals from rural areas, are familiar with woodfuels and their use, and perceive them to be safer than modern alternatives. Additionally, certain foods are considered to be better tasting when cooked over charcoal. In Tanzania, all of this holds true. Electricity and LPG are cheaper and more convenient household sources of cooking energy, yet even most wealthy people use at least some charcoal or firewood for cooking, for all of the previously mentioned reasons (Hosier and Kipondya 1993).

Urban Markets

The preceding discussion points to the unique nature of urban energy use. In order to feed the woodfuel energy needs of urban residents, a market for woodfuels has emerged in urban areas, with its own structure and problems, as will be described in this section.

Commercialization

The increasing urbanization of African countries has resulted in increased and greatly concentrated demand for and commodification of products, for which city dwellers must pay, rather than produce. These include agricultural products and woodfuels, and result in consequent development of systems to deliver these products to customers who demand it.

This commercialization of woodfuel has opened up a market that includes a large number of people and a great deal of money. In Zimbabwe, the commercial sale of woodfuels in the early 1980s had a turnover on the order of $ 90-100 million per annum (Katerere 1984). In Kenya, the fuelwood sector is estimated to employ over 2 million people (Leach and Mearns 1988). The traders who facilitate the movement of the commodity from the rural producers to the urban consumers dominate the market. These traders are the focus of much of the controversy over woodfuel markets. Their alleged business practices raise ire in consumers, politicians and researchers alike, and it is these traders who have traditionally been the target of attempts to reform the woodfuel system in urban areas.

Prices

To a large extent, prices define the urban woodfuel market, and greatly affect the impact that the market has on both rural and urban dwellers. The price paid for wood at its source will affect the price of woodfuel products for end users, the supply of woodfuel and the sustainability of wood resources. The price of woodfuel for urban consumers affects the standard of living of all users and the survival status of the most marginal user groups. In some low income households in Tanzania, for example, as much as 30 per cent of the household budget may be spent on the provision of energy (Hosier and Kipondya 1993). A rise in woodfuel prices may mean a change to cheaper fuels. But if there is no cheaper fuel available, the price rise can mean a reduction in use by the poorest people, which may translate into fewer meals cooked, or a change of food type to a less nutritious, but faster cooking food.

Although it has long been assumed that woodfuel prices rise in response to scarcity (deforestation) and increased transport distances, these trends have been seen to be less than straightforward. In fact, real prices of woodfuels in many cities have actually fallen over time, despite perceived and actual deforestation in the woodfuel catchment area for the cities, increased fuel transport distances and growing populations (Leach and Mearns 1988, Leach 1987). Short-term seasonal price fluctuations are influenced by supply issues, but the magnitude of the fluctuations (up to 100 per cent price increases) imply that there is more influencing prices than just supply swings (Leach and Mearns 1988).

Markets

If the price of woodfuels is only partly determined by supply and demand parameters, market imperfections and distortions may be the other important influence on prices. Markets and how they work can have a great influence on prices faced by end users. A market that does not work well will impose more costs on the trader, which will be passed onto or imposed directly on the consumer. Extra costs imposed by poorly working markets also affect the sustainability of the resource by encouraging traders and consumers to acquire fuel in the cheapest, and often least sustainable, way possible.

Market imperfections In the past, researchers and governments have worked with assumptions formed through anecdotal evidence, misunderstandings, and biases about the competitive nature of woodfuel markets. The popular view in most developing countries is that woodfuel markets do not work efficiently. Their deficits in price-signalling are held to allow monopolistic elements to exist in the system, and to permit large variance between prices paid by end users and those paid to producers and harvesters of fuels, differences which persist over time and space. What is not clear is whether these perceptions are, in fact, true in the case of the woodfuel market. Their verity or lack thereof is important in the formulation of policy. Even if the market is found to be generally efficient, there may be imperfections whose elimination would improve the sustainability or acceptability of the woodfuel market for consumers and market participants.

Another widely held belief is that Africans are uneconomic, non-entrepreneurial, and that their presumed inexperience puts them at a competitive disadvantage to those with whom they wish to trade.[3] A widely accepted concern is that this inexperience allows certain participants in the marketing chain, in our case producers and harvesters of fuelwood and small-scale retailers, to be exploited by wholesalers. This exploitation can take the form of collusion on prices paid to producers or prices charged to retailers, to individual price gouging by wholesalers resulting in excess profits for them.[4] A lack of bargaining power in the case of producers due to their economic marginality can result in similar abuses. A related belief is that seasonal and annual price fluctuations are the result of this excessive profiteering and monopolistic tendencies by wholesalers.

An additional impression is that the market chain is rendered inefficient by the presence of too many intermediaries. It is assumed that the extra margins and 'nonproductive' labour that goes into bulk-breaking adds unnecessary costs to the product. If this indeed is the case, why are not these 'unnecessary' middlemen just bypassed (Jones 1972)? In fact, some researchers have argued that a long market chain is an efficient use of available resources (Eicher and Baker 1982). Fears persist that available economies of scale are therefore not realized, also affecting the efficiency of the system.

There is concern that the distribution system, in its present efficient or inefficient form, is not capable of handling increases in volume that may be necessitated by increasing population and urbanization in most countries. The transportation network serving to distribute the fuels may not be adequate to expand in the long run, for example. The long-run stability of the distribution system may, in fact, be better served by an alternative market structure (Mercer and Soussan 1992).

Finally, markets in developing countries are plagued with infrastructural and social barriers that may inhibit the spread of information through one or more levels of a market. This lack can reduce the effectiveness of an otherwise fair and efficient market (Klitgaard 1991).

Importance of Markets as Price Signalling Mechanisms

There is little agreement on the status of woodfuel markets in developing countries. In fact, there is no real agreement on how markets of any sort, even agricultural markets, function in developing countries. Yet, there is agreement that markets do have influence on consumers, and that markets and their functioning may very well be the important link between prices and consumer and environmental welfare, in particular in the case of woodfuels. This is due to their function as price signalling mechanisms. When performing correctly, their ability to indicate, in at least a rudimentary way, the human and environmental costs of a particular commodity can be invaluable to maintaining the sustainability of the human and environmental systems.

Importance of externalities Prices in a well-functioning market should take into account externalities, as described in a previous section. These 'hidden costs' are crucial to resource maintenance - if they are not reflected, the resource will be used in a manner incompatible with its true value to society. The understanding of woodfuel markets will encourage such sustainable use, allow policies to be formulated addressing questions of price, and avoid the social costs associated with the problem.

Consumer Welfare

The end effect of market imperfections is a loss in the welfare of consumers. Rural and urban consumers and producers, as well as intermediate participants in the market are served inadequately by a poorly functioning market. There is a short-term welfare loss experienced by consumers who pay more than they should for the fuel and producers who are paid less than they should be for fuel due to market distortions. There is also a long term loss to rural dwellers and the society as a whole when resources are used in a non-sustainable and inefficient manner.

Additionally, equity questions can be better addressed through a well-functioning market. The division of wealth between rural and urban dwellers will be more equitably arranged, as the resource wealth of the rural areas is compensated by the monetary wealth of the urban areas.

Conflict between Needs of Consumers and Environment

Even if markets are studied and understood, and improvements made in market information and infrastructure, questions remain concerning the inherent conflict between consumers and the environment. As discussed earlier, if prices of standing woodfuel were increased to reflect environmental costs of the resource use, the price of fuel to consumers would probably rise. This makes it even more imperative to identify inefficiencies in the markets which deliver the woodfuel to consumers, because the better the market works, the less inefficiency and the lower the final price. This can offset in part the price increases needed to compensate for environmental impacts of the resource use, but prices will inevitably increase. And for people who are at the edge of survival with no viable energy alternatives, as are many in African cities, that increase may be too much.

Overview of Study

A Move towards 'Perfect' Markets

There are two major points that have been identified in the preceding discussion as being at the core of problems with woodfuel markets in developing countries. Both influence the price of woodfuels faced by urban end users.

First, the network that supplies woodfuels may hide inefficiencies and distortions that make the price of woodfuels higher than they need to be for end users. These problems (or their lack) need to be identified and corrected.

Second, the price of woodfuels is distorted by the absence of environmental externalities in the price of the fuel, causing prices generally to be lower than they should be. These externalities need to be internalized in the woodfuel price.

This study will concentrate on the first point, the problems with the supply networks of woodfuels. The second point will be considered again in the final chapters of the study, as policy measures to improve the supply system as a whole are considered, and the two problems are integrated to look at the supply system holistically.

The research will examine the supply markets for woodfuel in urban areas *via* a case study of the woodfuel markets in three urban areas of Tanzania. It will attempt to determine whether these markets are performing at their best, and if not, where problems exist. From these results, the study will recommend what changes can be made to improve the markets and, hence the sustainability of the energy system in Tanzania. It will examine the connection between social and economic factors in the system, and determine the effect of the market and changes in it on the people whose lives the woodfuel sector touches.

Chapter 2 looks at rural markets in developing countries. First, the importance of markets is discussed. An examination of previous knowledge about woodfuel and agricultural markets and a determination of where they overlap follow this. After identifying imperfections identified in previous studies, the chapter concludes with a statement of the goals of this research, followed by a description of how the study will be done.

The next chapter outlines the process of collecting data for the study. It describes the design of the survey, the selection of a sample, and the implementation of the survey in Tanzania. Included is a critique of the methodology used in the survey. This methodology was untried for a survey of this type, and the advantages and disadvantages of the methodology as compared to that used in previous surveys is analyzed.

Chapter 4 is the first of four chapters analyzing the data that was collected. Chapter 4 begins by analyzing the structure of the woodfuel market. It looks at demand and supply of woodfuel, the social and economic characteristics of participants in the market, and the motivations and requirements for participation in the market. Problems and successes of the market at this level are identified, and their potential effect on human and environmental welfare discussed.

The next chapter continues the analysis of the data, examining the conduct of market participants. It explores how sellers set prices and the social and economic relationships between buyers and sellers. Again, problems are identified, and their effects discussed.

Chapters 6 and 7 scrutinize the performance of the market. They look at market efficiency of different sorts, examine prices and profits at the time of the survey, historically, and between urban areas. As before, problems, successes, and effects of the market performance are discussed.

Finally, chapter 8 discusses what has been found by the analysis. There, the imperfections of the market as a whole are tabulated, and policies and infrastructure improvements needed to correct them are discussed. Also discussed are the social and environmental implications of reform and lack of reform.

Notes

1. Cf. Anderson 1986, Sharma 1992.
2. Examples include areas in Somalia and India (Mercer and Soussan 1992).
3. As they relate to subsistence food markets, these perceptions may be seen in, for example, Jolly 1989, Ellis 1981, Jones 1974, Ruttan 1969, and Eicher and Baker 1982. See following section for a comparison of food and woodfuel markets.
4. Examples of such claims for woodfuel markets include French 1984, and Mazambani 1984.

2 Woodfuel Markets

Importance of Markets

As discussed in some detail in the first chapter, woodfuel markets are an important piece of the puzzle that determines the effectiveness and sustainability of the use of wood as a major source of energy in African countries. Woodfuel markets influence the price of fuels in the marketplace, and so their use by consumers and their ability to survive affect the harvesting of trees in the forest, both in quantity and location. This ultimately affects the lives and livelihood of rural dwellers, and, in the long run, determines the ability of the country to meet the energy needs of its citizens. The effective functioning of woodfuel markets has the potential for touching the lives of a great majority of the people in a country that depends heavily on woodfuel for energy. It becomes important, knowing this, that the woodfuel markets in such countries are well understood. This understanding comes both in the isolation of the market as an economic entity, and in the context of the social, economic, and governmental systems that help to define the markets. From this understanding can come both specific and general measures that will help maintain the sustainability of energy supplies and forests in affected countries.

This chapter will introduce some definitions that will be invaluable in the analysis that follows, and will introduce the work that has been done by others to lead up to the present study. This work has been done in and out of Africa, and has looked both at woodfuel markets and problems and at similar rural-urban product distribution systems. Finally, the chapter will formally introduce the questions that the study proposes to answer, and present a framework for their analysis.

Markets

Markets and Market Chains

Woodfuel markets are defined in this study as the commercial distribution systems which spatially and economically link producers and consumers of woodfuel. The producers and consumers themselves are also included in the entirety of the woodfuel market. The woodfuel distribution system is made up of a number of stages, as the fuel is moved from growing stock in the forest to the cooking stove or other use in the cities. This progression of stages which is responsible for the distribution of woodfuels is referred to as the *market chain*.

Competitiveness and Efficiency in Markets

The concepts of markets, market chains, and the failure or effectiveness of market behaviour are economic in nature. As such, they must be defined in terms of economic theory. The basis is the idea of efficient markets. An efficient market is one that maximizes the welfare obtained from the market. In other words, an efficient market will get the most out of the set of resources available to it. In the same way, an inefficient market does not achieve the maximum possible benefits from a given set of inputs. There is consequently a loss to the economic system as a whole because, ideally, more benefit should have been extracted from the resources of the economy. As such, economists strive towards the ideal of a completely efficient market, fully realizing that it is an abstraction that is basically unachievable in the real world.

Another abstraction is used as the basis for determining the efficiency of a market. This is the model of perfect competition, which has been widely applied in developing countries to agricultural and other markets. Economists use it to study market behaviour. The model, in its most simple form, has the following assumptions (Klitgaard 1991 after Jones 1972).

- The largest firm makes up a small fraction of the total sales or purchases of the industry.
- Market participants act independently and impersonally.
- Entry to the market is free.
- The traded commodity is homogeneous and fungible.
- Only the participants in the transaction determine the time and conditions of sale.

- Participants are economically motivated.
- All market participants have complete knowledge of offers to buy and sell.

Traditionally, the model is used to trace price movements as a function of supply and demand. Prices that follow the competitive model signify an efficient market, those that do not indicate an inefficient market. Conversely, an efficient and well-functioning market should allow all of the above assumptions to be true, and any government intervention in such a market is inefficient (Klitgaard 1991). Conditions that veer from the listed assumptions are called market imperfections, and are a type of market failure.

Although a competitive market is necessary for an efficient market, it is not sufficient. Market imperfections are only one of several market failures that can be identified. Others include externalities and public goods, increasing returns, and distributional inequity (income and wealth) (Wolf 1988). Any of these failures will cause a market to be inefficient, and the potential for behaviour or conditions that would render the market inefficient exists at each stage of the market chain.

Price Controls in Tanzania

Never in recent memory has Tanzania had a political or economic commitment to the ideals of the 'free market' and the idea of a market clearing price for goods. Tanzania has had price controls in some form since 1920, and throughout the 1970s and 1980s, Tanzania had in place a comprehensive system of price controls (Whitworth 1982). Established in 1973, these controls were closely tied to the philosophy of *Ujamaa* or 'African Socialism' personified by Julias Nyrere (Bevan et al. 1987). They were a reaction to the serious inflation of the early 1970s, and a logical consequence of a wages, profits, rents and productivity policy enacted in 1967 (Semboja and Rugumisa 1988).

Over 1,100 prices were controlled in 1973, and the list of controlled prices continued to grow until at least 1978. Although meant to increase equity in the country, the controls went beyond the necessities of life to set the prices of such luxury goods as liquor, refrigerators, cigarettes, bicycles, cheese, etc. (Semboja and Rugumisa 1988). These price controls were pervasive and strictly enforced in the urban areas. They led to severe shortages and long lines, and, in reaction, a black market and informal sector to provide goods that could not be obtained officially (Bevan et al. 1987).

Interestingly, the prices of woodfuels were never controlled, despite their status as basic needs. The reason may centre on the fact that the woodfuel marketing system has always been a domestically focused, informal sector operation. Subsistence food marketing also fulfils those requirements, yet some 'essential' subsistence foodstuffs were subject to price controls. The prices of other marketed energy sources, all imported, were controlled. Most controlled items could be found only on the black market, and items which had prices that were not price controlled, including woodfuel, were the most reliable.

In 1986, the International Monetary Fund and the Tanzanian government agreed on an economic recovery program (ERP). Trade liberalization, the removal of subsidies, exchange rate adjustments, increases in maximum prices of officially controlled commodities and fewer price controlled items are all part of the ERP, and many price controls were removed or decontrolled by the early 1990s (Semboja and Rugumisa 1988). Although there is debate over whether the price controls should be maintained for true necessities (particularly food, electricity and water), the implementation of the ERP seems to have induced the government to give some latitude to the principles of the free market.

In many other developing countries a similar pattern of minimal regulation of the woodfuel market in the face of regulation of other markets in the country seems to have occurred. The study of woodfuel markets and pricing of woodfuel, as well as research on subsistence food markets, is discussed in the remainder of the chapter.

Markets in Developing Countries

Previous research on markets in developing countries comes from two areas. The first is studies that have looked at woodfuel systems and urban woodfuel markets in Tanzania, elsewhere in Africa, and in developing countries in other parts of the world. The second, more voluminous body of literature focuses on agricultural markets. Economists, geographers, anthropologists, public policy experts, and other social scientists did this work. The argument will be made that this literature, although it discusses the agricultural sector, is applicable in whole or part to woodfuel markets.

Woodfuel Pricing and Marketing Systems

Recently, research has focused on woodfuel pricing and marketing systems, primarily in Africa and Asia. Though many do not focus primarily on the distribution system in their studies, a few previous studies have looked at fuelwood marketing systems in Tanzania and other countries. All have found problems in the markets studied which detract from the efficiency of the market. The bulk have concluded in their studies that the woodfuel marketing system in the study area is competitive, but has one or more market failures which detract from the overall efficiency of the market. A smaller number have concluded that the market lacks competitiveness, which combines with other factors to make the woodfuel distribution system inefficient. A final inefficiency has been identified as government intervention in an otherwise well-functioning market.

Competitive but Inefficient Systems

Tanzania In a case study of Tanzanian cities based on the results of a 1987 World Bank survey and previous survey data, Leach and Mearns looked at Dar es Salaam and three other cities (Mwanza, Arusha and Dodoma) and delineated the pricing structure, intercity seasonal variations, and market characteristics of the woodfuel marketing system (Leach and Mearns 1988, World Bank 1987b). At the time of its implementation, the World Bank study was the most complete and, particularly in the area of market structure and trader margins and costs, the most detailed to date. However, it suffered from time constraints that kept its sample small and somewhat limited in scope. From that study, Leach and Mearns conclude that charcoal markets in the city are highly competitive, and that prices positively reflect economies of scale and vertical integration. They judge that the system is moving towards greater organization and integration, but that inefficiencies exist in the system, particularly in the transportation of the fuel, and acknowledge the need for further investigation into the details of the trading networks. While claiming competitiveness within the system, the Tanzanian case study gave no details about the reasons for this conclusion (Leach and Mearns 1988).

The Sahel A study of some Sahelian charcoal marketing systems describes the physical structure of the system of fuel provision (Bertrand 1985). It emphasizes transportation inefficiencies, as well as the decreasing margins of charcoal sellers, and increasing margins of producers over time. The study concludes that the data contradicts the 'myth' of fuelwood traders generating excessive incomes, due mostly to the effect of transport prices. It also notes that retail prices for woodfuel in some cities are rising slower than overall consumer price increases. All of these effects are attributed to the inability of traders to increase retail prices enough to completely compensate for increased transport charges as the cities' catchment areas expand and distances travelled by fuel grow.

India Outside of Africa, two studies in particular have thoroughly examined the workings of the woodfuel markets in urban areas of India (see Alam et al. 1984, a less detailed report of the same study, Alam et al. 1985, and Dunkerley and Gopi 1985). The authors concluded that the market was reasonably efficient and competitive at the retail level, but they were concerned that barriers to entry or other inefficiencies existed at the wholesale level. The authors also were concerned about the efficiency of the transportation system and the amounts of imported fuel used in transporting fuelwood products to the urban areas. They recommended greenbelt plantations around the urban areas and further investigation into improved charcoal kilns.

Haiti Finally, a study in Haiti designed to assess the impact of fuelwood use on deforestation describes the structure of the commercial trade of fuelwood in some detail, and assumes a fairly competitive market (Stevenson 1989). He discusses the efficiency of the system in terms of the number of members of the distribution chain, and finds he is able to explain differences in temporal, spatial and quantity prices by describing the value-added and risk components of wholesaler and retailer margins, debunking assumptions of usurious pricing.

Noncompetitive Systems

Senegal The situation generally assumed to hold true in African woodfuel markets by analysts of fuelwood issues is that of a highly competitive small-scale retail market supplied by an imperfectly competitive wholesale market (cf. Soussan et al. 1990, French 1984) The situation in Senegal is a well-documented representation of this view, but it is not known if this assumption holds true in most countries, or even in individual cities.

A study thoroughly investigated the imperfectly competitive atmosphere among wholesalers in Senegal (Ribot 1988). The government promulgated this situation when they installed a system of limited permits for harvesting wood and producing charcoal. The regulations also include the requirement that all charcoal merchants form into cooperatives that fix the retail price of charcoal. These regulations were meant to maintain the sustainability of woodfuel supplies, considered to be inadequate to meet the long-term needs of the country. Instead they have succeeded in drawing down supplies and concentrating economic power in the hands of influential merchants.

Zimbabwe In contrast, a study of Zimbabwe's woodfuel market makes largely unsubstantiated claims of usurious wholesaler profits and woodfuel shortages attributed to both transportation bottlenecks and vendor arbitrage (Mazambani 1984). The author blames the informal nature of the market and recommends government regulation to control prices, establish controlled access to retailers by licensed wholesalers, and create government run peri-urban woodlots. Unfortunately, the author interviewed only vendors and wholesalers who possessed one of the restricted licenses and who resented the competition of unlicensed participants.

Dar es Salaam Another study of the Dar es Salaam charcoal production and marketing system was done by the Tanzania Industrial Studies and Consulting Organization (TISCO) in 1986 (TISCO 1986). It identified the ad hoc organization of the producers and sellers of charcoal; the absence of large wholesale depots to ease the distribution of the fuel within the city; and the high cost of the transportation and production of charcoal as problems in the charcoal distribution system. The suggested remedy for these problems was the formation of fully integrated village charcoal production and distribution cooperatives. It also made implicit and unsupported assumptions about the noncompetitive nature of the woodfuel traders, and the resulting upward pressure on prices, and its recommendations reflect those assumptions. A small sample hindered this exploratory study.

Effects of Government Intervention

As an added note, some studies have found, when examining the market structure of the woodfuel distribution system, that the systems are noncompetitive due to direct government intervention. Often, these controls have turned a working, if not entirely competitive, woodfuel supply system into a less effective and inefficient system. The previously cited study on Senegal (Ribot 1988), a study looking at fuelwood use in Raipur, India (Macauley et al. 1989), and a paper looking at the Ethiopian woodfuel distribution system (Ellis 1988) all observed that despite close government control over fuelwood supplies and the nationalization of the fuelwood trade, private, generally illegal, trade flourishes, and actually provides much of the woodfuel supply in the urban area. Before the controls, the urban market provided adequate incentives to support a 'productively organized' entrepreneurial supply system, even if the market could not be described as completely competitive (Ellis 1988). Government interventions have alienated villagers in the supply areas and exacerbated price increases for fuel in the countries in the three cities. This apparent sidestepping of government control of the supply system is a telling example of the futility of government control of these informal markets.

Market Imperfections

The results of this set of studies are difficult to generalize. Not only did each study look at the organization and efficiency of the distribution system of woodfuel in a slightly different way (and many only as part of a broader analysis of woodfuel use or supply), each also found different problems or strengths in the local system. Although the problems most often identified by the previous examinations of woodfuel supply systems are transport inefficiency and high cost, and inefficiencies or usurious practices at the wholesale level, these results were not universal. In fact, the studies which were more rigorous in their analyses of the systems (Bertrand 1986, Stevenson 1989, Alam et al. 1984 and 1985, Dunkerley and Gopi 1985, Ribot 1988) were much more likely to find the woodfuel supply system generally competitive and efficient, and less likely to find that wholesalers generated usurious profits or engaged in collusive practices. They might, however, identify some inefficiency in the system at the wholesale or other levels.

Need to go Further

The investigations of woodfuel market efficiencies have generally concentrated on historic and cross sectional prices and margin analyses. The analysis of margins and price comparison are important tools in the inquiry into the efficiency of a marketing system and give an important first cut impression of potential problems in a system. However, they do not delve deeply and systematically enough into the interactions between market structure, market participants, and the prices of woodfuel at various stages of the trip between production and end use. It is this more comprehensive type of study which is lacking, and which is presented in this dissertation.

Similarities between Woodfuel and Staple Food Crop Markets

Most of the imperfections in the market described in the section above are generally associated with informal markets of all kinds. These include the woodfuel market and the subsistence food market. Studies looking at imperfections in informal markets have most often been concerned with staple food production. The woodfuel and subsistence food markets are alike in many ways, in their structure and scale as well as in their ties with rural agriculture, and parallels can be drawn between research in the two areas. They also share similarities with rural industries of other types. There are, however, key differences which make the present work vital for planning in the woodfuel sector.

The primary distinguishing features shared by all rural industries are the characteristics of a rural production base providing goods for urban consumption and the small scale of their production and distribution systems. All share a need for a distribution system that moves the rurally produced goods into the towns for consumption by town dwellers. In most cases, this distribution system is based in the private sector, and subsists on private capital. Food and woodfuel supply systems differ from other rural industries in that their production base is more widely spatially dispersed. This is particularly true for the woodfuel supply sector, where at times woodfuel is harvested far from populated areas. In many cases, however, woodfuel harvesters and charcoal producers work as full-time farmers, and sometimes the land from which they harvest the woodfuel is later used for agricultural purposes.

In contrast to the majority of staple food crops, the bulk of woodfuel products tends to prevent them from being transported great distances. As such, it is not expected that the supply areas for woodfuels would be as extensive as they are for staple food crops. This is due in particular to the fact that woodfuel products are more homogeneous than food crops; i.e. there is far less difference between the charcoal from one location and another than between, say, millet and corn.

The seasonality and year-to-year fluctuations of agricultural production distinguishes it from other rural industries, including woodfuel. Although woodfuel production is often seasonal, the seasonality is caused by the time available to producers (farmers have little spare time in the wet season) and infrastructural conditions (roads are poorer in the wet season) rather than natural constraints that are embodied in subsistence agriculture. This implies that temporal fluctuations in woodfuel production may have different causes and solutions than those of subsistence agriculture.

An additional feature shared by woodfuel and staple food crop supply systems is that they are primarily domestically based. Although some staple crops are also exported, and in some cases charcoal is exported, in general the staple food crop and woodfuel markets are confined to the country in which they are produced.

The two products are alike in that they exist outside the cash economy for some urban and many peri-urban and town dwellers, who grow all or part of their food and collect all or part of their woodfuel. Urban markets for the two types of commodities are similar as well. Both cater primarily to households and informal sector industry, and both are generally sold by small retailers and in small quantities to these consumers. Sellers for woodfuel and staple food crops are both found often in urban marketplaces, and the type of seller-buyer relationships may be similar.

A World Bank report described the agricultural market in developing countries as follows:

> Farm sectors comprise highly competitive businessmen, producing a relatively homogenous commodity for sale in a market with numerous, price and quality conscious customers. In other words, agriculture would appear to be the ideal industry in which to realize a textbook-perfect competitive market to the benefit of both producers and consumers (Klitgaard 1991).

These qualities are shared by the woodfuel sector, as is evident from the preceding section. It implies that woodfuel market, too, can be looked at under the rubus of the competitive market model.

These similarities describe parallel system of commodities and marketing systems which are alike enough to allow us to use some of the frameworks and assumptions found in numerous studies of staple food production and marketing as a starting point for the under-explored woodfuel marketing systems. The differences between the two commodities' production and marketing systems must be kept in mind when using such comparisons.

Staple Food Markets

Anthropologic Studies

Personal bonds The first and earliest studies of food markets in developing countries are anthropological studies of the 'bazaar economy', a term popularized by the work of Geertz (1978) in the 1970s. These economic anthropologists focused mainly on the price setting mechanism used by retailers in the informal market setting. They investigated the information component of the bargaining relationship between buyers and sellers, and later related the bargaining process to the efficiency and competitiveness of markets. Some of the earliest studies looked at the personal bonds that are developed between trade partners at all levels of the marketing system, and sought to explain their existence as being a risk averting strategy that maximizes efficiency in the short-run (cf. Mintz 1961, Davis 1973, Geertz 1978, and more recently, Plattner 1983, Alexander and Alexander 1987). Their work helped to address and discredit the Western assumption that people in developing countries are non-economic and irrational. Later works have described the connection as primarily economic, existing to overcome economic constraints (Trager 1988). Many studies nonetheless describe the small retail vendor as a marginal player, basically eking out a subsistence existence and looking not for expanding profits, but basic security, and also using the market as a vehicle to fulfil other social needs (cf. Szanton 1972). They point out that many retailers have multiple selling goals, and that profit is often not the highest-ranking goal (Jolly 1989).

Goran Hyden (1980) looked directly at Tanzania and described an economy and mode of production that he believes to be dominant in Tanzania, in Africa, and perhaps in much of the developing world. He calls it the economy of affection, and describes it as the natural outcome of a peasant mode of production. This is parallel to the market economy, which evolves from the capitalist mode of production.

The economy of affection is based on familial and communal ties, which are also the basis for organized activity (Hyden 1980). The basis of the economy of affection is reproduction, rather than production, and economic action is marked not only by individual profit, but also by social considerations. Maintenance of position in community, kinship and religious networks can return more and reduce risk to a greater extent than a profit motive will allow (Hyden 1980). This mixture of the market economy and the economy of affection affects the motivations and actions of traders and others involved in the modern economy. It is the basis of the series of personal bonds that are described in the studies considered previously.

Subsistence food markets as models of pure competition. A subset of anthropological and cross-referenced economic studies looks specifically at the competitiveness and efficiency of markets. One early effort, a detailed case study of wholesalers in a Jamaican food market, finds the market in a state of pure competition, and that the system is efficient, though labour intensive (Katzin 1960). This penchant for using subsistence markets as a model for pure competition is echoed in many other early studies (including, e.g., Belshaw 1965, Geertz 1978). Later work, however, refutes the idea of a market operating under pure competition (Norvell and Thompson 1968). This later study remarks that the price inflexibility described obviates the pure competition model, and argues that the model additionally is a poor test for allocative and pricing efficiency and does not take into account the spatial aspects of uncertainty.

Other studies emphasize the fluctuating competitive and structural atmosphere in a market, and suggest that in a working market there is a continuously changing price structure based on these fluctuations (Goldin 1986, Jolly 1989). Many of the studies emphasize the lack of available credit as one of the most restrictive aspects of market competitiveness. One of the few studies in this genre that looks at the national and wholesale markets agrees with this analysis. It finds that competition is severely restricted in the Burkina Faso grain market, mostly due to concentrated control over seasonal storage of grain because of lack of available credit (Saul 1987). It also cites collusion between wholesalers in some villages, and concludes that,

> in villages, islands of controlled competition and oligopsony are as much evident as those of pure competition, and that it would be wise not to jump to assumptions of structural conditions for perfect competition from the observation of 'hundreds of small traders' at the national level (Saul 1987, p. 80).

Market imperfections This genre of geographical and anthropological studies of food markets points out that competitive and economic food markets may exist in developing countries, particularly at the retail level, but that they may operate under conditions that do not meet the classical models of pure competition. In particular, the existence of personal bonds, sometimes substituting for otherwise inadequate sources of credit, is shown to be economically rational for participants. At the same time, personal bonds can be the source of large overall system inefficiency. They point out the necessity for studies that look beyond a strict economic model and take into account the social and political influences that are inevitable in any marketing system, and which seem to have particular influence in food markets. Saul's research also indicates a need for studies which look at competitiveness at every level along the path of a marketing system, as opposed to concentrating only on retailers or wholesalers, for example. The studies are particularly good at pointing out the social aspects of markets, but have traditionally been reluctant to link these factors with the economic structure of the markets.

This flaw has been addressed in other studies. One study, for example, focused on Dar es Salaam, the capital of Tanzania (Lynch 1994). This study examined the interaction between the formal and informal agricultural marketing systems and found that the informal system has had the flexibility and dynamism to physically and economically adjust to changing economic, political and social conditions.

Economic Studies

A second, and closely linked, type of market research looks particularly at the economic structure of food markets. This research was done by agricultural and marketing economists and geographers looking at subsistence agricultural markets in Africa and South Asia starting in the mid-1960s and lasting through the 1970s. The studies examined the marketing cost of food from producer to urban households, and explored imperfections and inefficiencies that constrain competition in the markets. These are spearheaded by the work in Africa of William Jones of the Stanford Food Research Institute, who based his research partly on work done earlier in South Asia (see Jones 1972, Lele 1971). Jones pioneered the use of the structure, conduct, performance (SCP) paradigm, borrowed from industrial organization theory, in studies of this kind. He uses this revised neoclassical economic methodology to explore the allocative efficiency of agricultural markets, and to determine the overall competitiveness of the system, and in so doing challenges the validity of many previous studies reporting market imperfections (Eicher and Baker 1982, Miracle 1968).

India Lele (1971) used a similar methodology in her study of Indian food markets. She refutes contentions that private agricultural markets in developing countries are not highly competitive. By direct observation of markets she finds them to be open and competitive, and by analyzing inter-regional and temporal price variations she finds regional price differences can be attributed to transport bottlenecks, but are otherwise indicative of a reasonably competitive system.

Africa Jones' (1972, 1974, 1987) analysis spans several African countries, but focuses mostly on Nigeria. It tests the efficiency of the market in three ways. First, the extent of market integration was tested using correlation of market prices. Second, spatial relationships were examined by comparing prices between markets in relation to transportation costs. Third, storage costs were compared with seasonal price changes in order to determine whether prices were temporally linked, and temporal price increases not caused by excess profits for middlemen. He found that marketing chains were relatively short, and markets were only weakly integrated. He also found that poor physical infrastructure did not restrict market competition, that entry into markets was adequately free, and that there was little evidence of collusion and exploitation of farmers and retailers by middlemen.

Similar findings, supporting the conclusion of reasonably efficient and competitive markets and sometimes their weak integration, have been reported by other researchers (cf. Southworth et al. 1979, Berg 1980, CILSS/Club du Sahel 1977, Hays 1975, Hays and McCoy 1978, Raju and Bhatt 1985, Sherman 1985, Hollier 1986, Epstein 1982, Delgado 1986, Schmidt 1982).

Criticism These findings and the SCP methodology were sharply criticized by Harriss (1979, 1982). She claimed that correlation coefficients are not an appropriate measure for either market integration or competition, and that the data used in determining them were generally weak. Additionally, she criticized the assumptions made for spatial and temporal analysis as being overly simplifying, contended that most of the analyses have shown 'a serious lack of logical relationship between the data presented and the conclusions derived', and claimed an ideological, anti-interventionist motive for the fixation on competition in such research. Her own studies and her recommendations in this criticism stress the need for more research on structural relationship between production, exchange, and consumption (Harriss 1981, 1984).

Market imperfections The economic analyses of market performance using inter-market and inter-seasonal price movements show the following market imperfections: poor information, which results in high transaction costs; poor transportation infrastructure resulting in high handling costs; unstable supply channels due to lack of control over weather, disease, and storage losses; variation in product quality; but a seeming lack of collusive practices, cartels, and monopsonistic practices putting pressure on producer prices (Eicher and Baker 1982, Barrett 1997, Jayne and Jones, 1997). Eicher and Baker contend that research based on market imperfections and efficiencies has 'frequently generated few policy relevant conclusions because it has relied too much on idealized models' (p. 187). Nonetheless, the results and methods of such studies are treated as the standard in many overview publications (cf. Timmer et al.

1983, Wortman and Cummings 1978). The lack of alternative methodologies makes this inevitable. A merging of the anthropological type of study focusing on social systems and their influence on marketing systems with the more strictly economic emphasis of the studies reviewed here would contribute greatly, and would more closely emulate the real world situation in which the markets exist.

Findings on Competitiveness and Efficiency

Looking at the body of studies relevant to the issue of competitiveness and efficiency in woodfuel supply systems, it is clear that there is a need for studies both providing data and analyzing existing data in a manner which specifically addresses the extent of efficiency and competitiveness in the marketing systems and the effects of existing inefficiencies. Though there is a base of analytical studies on food marketing extending back for three or more decades, neither the approach nor the knowledge gained from these studies has ever been applied to woodfuel markets. While the topic of market competitiveness comes up at times in the woodfuel literature, it is often the subject of unsupported assumptions. When market competitiveness and efficiency is a subject of some numerical analysis, the analysis stops short of a full, in-depth investigation into the market interactions and price reactions. Because of this, it is difficult to make any generalizations concerning the competitiveness and efficiency, and, in effect, the sustainability of woodfuel markets in developing countries as a whole or in individual countries. Prevalent myths of the predominance and unfair advantages taken by middlemen are also unable to be proven or disproved, and so persist, unsubstantiated.

The SCP paradigm as developed for studies of food markets in developing countries has provided in that area a framework for the sort of in-depth analysis that is lacking in the woodfuel literature. Although the methodology has been criticized, it is widely accepted as the leading framework for investigations into the kind of domestic distribution system that characterizes stable food crop markets. In consequence, the paradigm can be an effective framework to fully explore the efficiency and competitiveness of woodfuel markets and their distribution systems.

Many of the studies in this genre, both those investigating food markets and those looking at woodfuel markets, show awareness of a need for studies blending economic and social analysis, though many are not successful in achieving this blend. Although based on an economic paradigm, the SCP methodology as developed for explorations into food markets has been modified to take social interactions in the market into account. The methodology has been rightly criticized for still not being sensitive enough to non-economic market forces, and so it will benefit from further modification and addition to better reflect these concerns when it is applied to woodfuel markets.

Additionally, these studies have said little about the influence of markets on the sustainability of systems. Although the sustainability of economic systems is an implied goal of economists looking at market systems, the more direct goal of maintenance of woodfuel markets because they are a necessary part of the social and economic system of the country is rarely emphasized. As well, the added emphasis on non-economic factors will improve the responsiveness of the research to sustainability goals.

The Research

This work aims to look at woodfuel markets as the determining factor in the sustainability of woodfuel in developing countries as a source of energy for urban households and the standards of living of these families and their rural counterparts. The environmental effects of woodfuel use will also be examined.

The initial goal will be to determine the level of efficiency and competitiveness in the woodfuel markets in Tanzania, looking especially the temporal and spatial pricing strategies of the participants in the markets, from producer to end user. From the results achieved through this analysis, the questions of sustainability will be discussed.

Structure, Conduct, Performance

The first step in the procedure is the gathering of data. Because there is little data on urban energy use, preferences, or the distributional system in Tanzania, an extensive data collection regimen was undertaken. The details of the data and their collection can be found in Chapter 3.

The procedure followed in analyzing the gathered data to determine the competitiveness and, ultimately, the efficiency of the woodfuel market structure in Tanzania is loosely based on the standard industrial organization structure, conduct, performance, or SCP, paradigm developed by Bain (1959).

The goal of the SCP paradigm is to trace the existence or absence of significant imperfections in the market, thereby identifying conditions departing from the conditions of a perfect market. From this information, the primary goal of answering underlying questions about the costs and efficiency of marketing services and the ability of the system to transfer information about the resources and long-run structural transformations associated with the system can be achieved. Though the idealization of a perfect market is not directly transferable to the real world, it provides a basis for analysis.

A constraint on this type of analysis is the lack of complete and reliable data, and though the data collected here are of good quality, and the author was personally involved in its collection, there are nonetheless holes that can not be closed. As such, an indirect approach based on the normative competitive model described earlier in this chapter is followed. In other words, it is assumed that, for example, a well-functioning woodfuel market *should* allow free entry, free moving prices, etc. (Klitgaard 1992).

This approach, based on Bains' original ideas, goes a somewhat different route. It goes beyond the economic considerations, and takes into account social and, to a lesser extent, governmental factors in the analysis. In this same way, the analysis also moves beyond Jones' (1972) look at staple food crop marketing.

Structure

The structure component of the SCP framework refers to 'characteristics of the organization of a market that seem to influence strategically the nature of competition and pricing within the market' (Bain 1959, p. 7). Under this umbrella lies questions concerning the existence of elements of a competitive market. These elements, fulfilment of which is sufficient but not necessary for a competitive market, include the following conditions:

- items of the traded commodity are fungible (interchangeable) and divisible;
- buyers and sellers act in an economically rational fashion (they want more, not less, incomes and goods);
- firms are small and numerous enough that their decisions have no impact on prices;
- all participants have equal access to activities of the market on the same terms;
- everyone has complete knowledge of forces likely to influence supply and demand (Timmer et al. 1983, p. 165).

This analysis will go beyond the framework set up by Bains. Specifically, the analysis will look at the seller concentration and the length and personnel make-up of the marketing chain by describing each level, tallying the number of levels and the number of buyers and sellers at each level, and determining the relative size distribution of each. The relationships between participants at different levels of the marketing chain, including the bargaining positions of farmers will be examined using self-reported responses, as will the entrepreneurial characteristics of participants. The barriers to entry at all levels of the system will be looked at by comparing the margins of participants at each level with other investments or jobs, the existence of license and entry fees, the availability of credit, and the influence of seasonality on market relationships. Finally, the extent of agent and product differentiation and the distribution of information will be discussed. The existence of low barriers to entry, lack of collusion, combined with reasonably accessible market information 'carries a strong presumption of a competitively efficient marketing system' (Timmer et al. 1983).

Conduct

The market conduct portion of the SCP paradigm is described by Bain (1959, p. 9) as the 'pattern of behavior which enterprises follow in adapting or adjusting to the markets in which they sell (or buy).' Again, the analysis conducted here goes beyond this basic model to include a host of social factors. In particular, an analysis of market conduct includes analysis of the methods of price determination, including communication and collusion within the market and the extension of credit as an exclusionary tactic. Additionally, this part of the analysis examines the attitudes of market participants towards sales promotion and competition with their fellow traders. All of these measures are qualitative and are based on survey responses and observation by the author.

Performance

Market performance refers to the economic and strategic end results of structure and conduct, and the adjustments engaged in by buyers and sellers (Boyer and Davis 1990 after Bain 1959). In the analysis, performance is broken into two sub-categories, productive efficiency and pricing efficiency. Under the productive efficiency rubric, allocative efficiency is analyzed by looking at capacity utilization, while scale efficiency is considered in two ways: first, by comparing the margins of small and large players at each level of the market system; and secondly by comparing the input and output of these players.

Pricing efficiency is looked at directly *via* the measurement of marketing costs and margins, and indirectly by analyzing the spatial and temporal components of price data. The analysis of average margins will look at gross marketing margins and net marketing margins and compare the two, compare wages earned to alternate jobs and investments, and look at returns to labour and, in some cases, investment. These measures will be used to determine whether monopolistic elements are present and whether excess profits are being earned, resulting in inefficiencies.

As an indirect attempt to determine the same facts, the spatial relationship between transport costs and intermarket price differences will be examined at the inter- and intraregional levels to test the effectiveness of merchant arbitrage over space and the interrelationships of the markets. Additionally, correlation between prices in the three towns studied, and also between markets within Dar es Salaam will be calculated to further address this issue, though, because of the structure of the woodfuels market the correlation is expected to be weak. The examination of spatial considerations will be concluded with a discussion of overlapping zones of competitiveness.

Temporal connections should also exist between markets, and therefore the study will examine seasonal and annual price movements. Marketing efficiency should produce seasonal price rises that correspond to the costs of storage. A seasonal index will be determined for each city and compared between cities and with the costs of storage. Respondents' qualitative opinions on the desirability of fuel storage will also be examined. Historical real price trends will be the next item examined, and the results compared between cities. An efficient system will reflect supply and demand changes in its price trend, but will be otherwise neutral. These trends will be looked for, and any deviations analyzed.

Results

After gathering the results of this series of qualitative and quantitative tests, the inefficiencies and competitive constraints on the fuelwood marketing system in Tanzania will be able to be analyzed. From these results the applicability of government intervention in the markets can be analyzed, and policy recommendations for efficiency improvements made. Even if the market is found to be efficiently allocating resources within the technical development parameters in the country, there may be much room for improvement of the system in the allocation of supplies or in reduction in infrastructure costs and bottlenecks. Lowering infrastructural costs can expand marketing capacity and improve market performance by lowering unit costs, regardless of the existing efficiency of the market (Southworth et al. 1979).

With this information, the sustainability questions posed earlier can be answered. Using the results of the analysis, particularly those pertaining to the efficiency of the market, the effectiveness of present policies towards maintaining market sustainability and the potential for new policy can be discussed. These qualitative measures are important to bring out the social aspects of the quantitative analysis, and to fulfil the promise of a study that goes beyond the monodimensionality of some previous research.

3 Data Collection

Data is always a problem in developing countries. The collection of data is time-consuming, expensive, and requires a level of infrastructural organization that is missing in most developing countries. In this regard, Tanzania is little different from its peers.

The existing urban energy and woodfuel market surveys in Tanzania were described in chapter 2. Overall, there existed previously a serious lack of information on energy use in the urban areas of Tanzania, and an even greater lack of data describing the structure and workings of the woodfuel distribution system. To address this lack of information the Tanzania Urban Energy Project was conducted in 1990. As part of this project, the surveys that form the empirical foundation for this research were undertaken. The methodology adopted for this research was developed to examine the spatial, economic and personal links between providers and users of woodfuel in Tanzania. The methodology was adopted in part to compensate for shortcomings evident in previous survey work done on energy sector issues.

This chapter will begin by giving an overview of the Tanzania Urban Energy Project. Then it will detail the collection of data for this research, the methods used and their advantages and disadvantages in comparison to work done previously, and the statistical make-up of the sample.

The Tanzania Urban Energy Project

The Tanzania Urban Energy Project was undertaken to gather data about the use of energy in urban areas in Tanzania. The project explored energy use in households, the informal sector, the commercial sector, the industrial sector, and the transportation sector. Additionally, the examination of the woodfuel distribution system that constitutes this research was undertaken as part of the project.[1]

For this research, the relevant parts of the project were the survey gathering data on the household sector. A household survey to collect information on energy use in the urban areas of Tanzania was conducted in July through

September 1990. Surveys were conducted in three urban areas, selected to represent a primate city (Dar es Salaam, the national capital), a secondary city (Mbeya, a regional capital) and a tertiary city (Shinyanga, a small regional capital and trading town). Figure 3.1 shows these cities and the main roads and railroads in Tanzania.

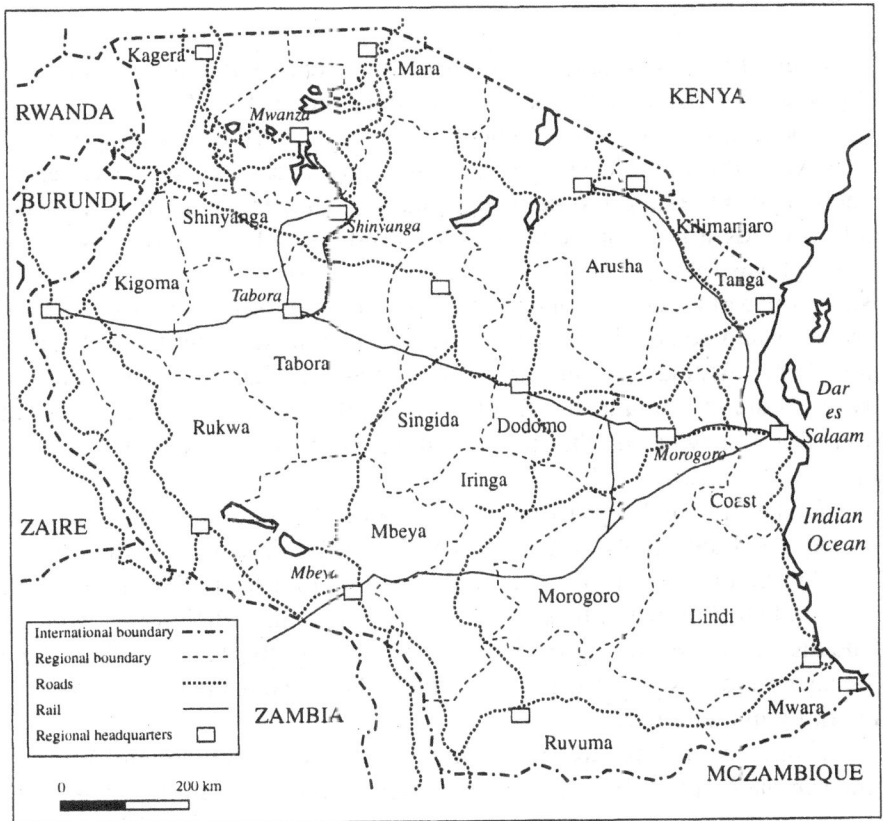

Figure 3.1 Tanzania

The household survey sample was based on a sample drawn from the census-sampling frame for each of the cities. The survey sampled 80 income stratified enumeration areas in Dar es Salaam, 31 in Mbeya and 30 non-stratified areas in Shinyanga.[2] A total of 2670 households were randomly chosen from within the sampling frame and interviewed. There were 1600 interviews conducted in Dar es Salaam, 620 in Mbeya, and 450 in Shinyanga. Households were asked about their fuel use, finances, attitudes towards fuel use, and the composition of the household. These surveys were used as a basis for the woodfuel market surveys conducted in September and October of that year.

Market Surveys

A series of surveys examining the structure, functioning and magnitude of the woodfuel market network were undertaken in the fall of 1990. A different approach was taken in carrying out this survey than is the standard practice in surveys of this type. Instead of surveying random samples of participants at each stage of the supply system (e.g., end user, retailer, wholesaler, producer), a backwards-linkage approach was followed, whereby the path travelled by the fuel was followed backwards from the end user to the producer. At each stage, the participant was interviewed, providing information about his or her business and personal characteristics, and specifically identifying his or her suppliers.[3] Although variants of this methodology have been used in other types of survey situations (cf. Jones 1974), this is the first time it has been linked in this way, and the first time it has been tried for surveys of energy market systems.

Under the usual approach a listing is attempted of each participant group in the woodfuel supply system (cf. ESMAP 1990, Openshaw 1989a, 1989b, 1989c, Alam et al. 1984, Dunkerley and Gopi 1983). This is generally accomplished by using lists of licensed retailers and wholesalers from city and district tax records, or from the licensing authority. At the producer level, licensing records are also used, or village heads consulted as to producers working in their areas.

From these listings are drawn samples, and interviews are conducted with the sampled participants. There are obvious problems with this method, the most glaring being the lack of completeness of listings garnered from official records. In all countries, and most especially in countries like Tanzania that have a flourishing informal sector, many sellers and producers work outside the official economy, and hence do not hold licenses of any kind. Because they are not listed officially, they are not sampled, and a significant bias enters into the sample survey. This problem is sometimes circumvented by doing listings of vendors encountered in known markets, and wholesalers spotted in wholesale markets or at roadblocks. In another survey technique, roadblocks are set up where all woodfuel-carrying vehicles or individuals are stopped and interviewed or counted during a fixed period. From these surveys, estimates of supplies of woodfuel entering a particular urban area, as well as characteristics of wholesalers and transporters involved in the woodfuel market are acquired. Unfortunately, transporters are known to evade roadblocks in order to avoid paying fees collected there, and so again some participants are missed.

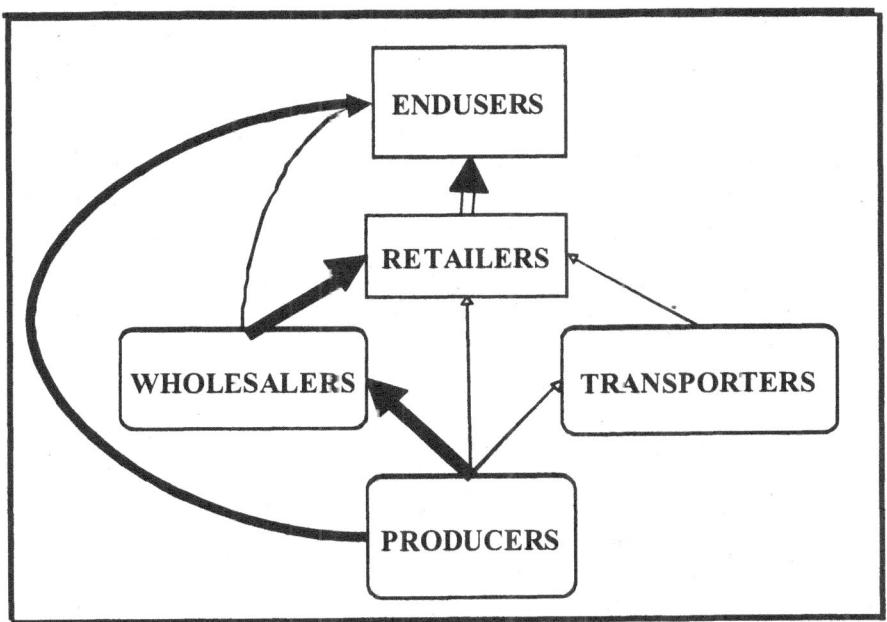

Figure 3.2 Woodfuel supply path – thicker lines indicate more commonly used connections

Backwards-linkage Survey Methodology

The 'backwards-linkage' survey methodology was developed to circumvent some of these problems. Instead of trying, and inevitably failing, to list all of the participants at each level of the woodfuel supply system and taking the survey sample from this list, the survey sample was taken from the existing and well-defined sample of household woodfuel users in an urban area. Four different surveys were conducted: an end user survey, a trader survey, a transporter survey, and a producer survey. They were designed so that each owner of a particular piece of fuel, from when it is harvested to when it used, was interviewed. The surveys were conducted sequentially, starting with the end user and working back to the producer, shows a simplified picture of the supply path. Much of the time, there were two levels of traders or sellers because both a retailer and a wholesaler (and sometimes even two wholesalers) owned the fuel as it moved from producer to end user. At other times, there were no intermediaries involved, only an end user and the producer from which the fuel was bought. One of the strengths of this methodology is that the incidence of this type of fuel path could be determined from the linked surveys, something that can not easily be done when unlinked surveys are used.

Each survey was developed especially for the research and for the methodology used. The survey instruments are reproduced in Appendix A. All surveys were conducted in Kiswahili, the national language of Tanzania, spoken fluently by all but the most rural and uneducated Tanzanians. The Kiswahili survey was translated from an English-language base survey and tested in Kiswahili. Rigorous training and monitoring were employed to ensure consistency and verity. The following sections will describe each of the surveys.

End user Survey

The first survey conducted was the end user survey. A sub-sample of end users was chosen from the sample of woodfuel users surveyed in the project's household survey done in the summer of 1990. Each household that used firewood or charcoal was considered to be one end user. In Dar es Salaam and Mbeya a sample of two users of charcoal and two of firewood (a 10 per cent sub-sample) from each of the enumeration areas of the original survey were randomly chosen. In Shinyanga a sample of one user of each of the fuels (a 7 per cent sub-sample) was chosen. At times, there were fewer than the one or two end users of a fuel in a particular enumeration area, and in these cases all of them were surveyed. The total numbers of interviews conducted is displayed in Table 3.1.

The end users were interviewed briefly to determine their buying habits and perceptions of supply and choice of fuel, and produced for every end user a list of one or more suppliers and the proportion of fuel bought from each of them. Then, using the proportions as weights, a supplier was randomly chosen. In this way, for every end user the chance of choosing a particular supplier was equal to the chance of this end user to purchase fuel form that trader. This ensured that all potential suppliers, even those that are used rarely, had a chance of being interviewed. It also ensured that, when weighted, the proportion of each type of supplier providing fuel to households in the city as found in the survey reflects the true proportions. When specific suppliers

Table 3.1 Survey sample size

	Survey type (number of people interviewed)			
	End user	Trader	Transporter	Producer
Dar es Salaam	142	157	20	99
Mbeya	120	103	6	47
Shinyanga	92	50	0	34
Total	354	310	26	180

could not be identified (for example, when an end user indicated that she bought from any of several sellers at a particular location) the location was recorded and a seller at that location was chosen at random.

It was not necessary to ask end users more than their suppliers, and the prices they paid for fuel, since a complete energy usage and attitude survey had just been completed by the same households.

Trader Survey

The supplier identified from the end user survey was interviewed in detail. Traders were asked about their social characteristics; their relationship with buyers and other sellers; and economic details of their businesses like costs, selling and buying prices, and other details. They were also asked to identify their suppliers and the proportion of their product they got from each of them, as in the end user survey. From the responses to the questions on suppliers, as above, the next node on the fuel path was identified. Again, this allowed potential suppliers of all types to have a chance of being chosen in direct proportion to their actual incidence in the population of suppliers.

The next seller in the supply path was often another woodfuel trader. If so, he or she was interviewed, as above, and the next supplier identified. Eventually another level of ownership in the fuelwood path was reached, either a transporter or a producer.

Transporter Survey

Very few transporters were interviewed. In studies that were done previously, from which an initial view of the woodfuel market was formed to aid in designing this research, transporters were identified as being part of the ownership chain of the woodfuel. However, as the survey work for this study progressed, it became evident that, although hired transporters carried most fuel, they rarely took possession of the fuel. Instead, they were just a cost incurred by the trader or end user. Since only owners of the fuel were considered for interviewing, very few transporters were interviewed. The exceptions tended to be transporters who were drivers hired to transport a different product who were moving charcoal as a sideline without their employers knowledge. These individuals tended to be reluctant to be interviewed, since what they were doing was technically illegal and could cause them to be fired from their job.

Producer Survey

The selection and interview process continued until a charcoal producer or woodcutter was identified and interviewed, thus completing the backtracking of the path of the fuel purchased by the original end user, and defining a set of spatially defined decision making nodes stretching from wood source to end use.

During the realization of the survey, all of the supply areas for each of the three cities were visited and producers were located and interviewed. In Dar es Salaam, the path from each household end user was followed to a producer, who was then interviewed. In Mbeya and Shinyanga, time and budget constraints allowed only one-half to two-thirds of the paths to be followed to their end. Nonetheless, enough information was gathered to get a good picture of the details of the producers and their activities.

Producers were asked questions similar to those asked of sellers covering economic, social, attitude and opinion issues.

Not infrequently, traders or end users would be unable to identify a particular seller from whom they habitually purchased their fuel. Instead, there was a particular market or village or roadside where they went to purchase woodfuel, buying from whomever was there when they arrived. In these cases, the interviewee was questioned to determine if there was someone at that location from whom they mostly bought fuel. If they insisted that they bought from anyone at a particular location, and had no particular ties to one seller, the location was noted and a seller chosen at random upon arriving at that location.

Advantages of the Backwards-linkage Methodology

This backwards-linkage methodology has several advantages *vis a vis* the standard methodology. As detailed above, in the previously used methodology, a sample of participants, such as wholesalers, are chosen from what is almost always an unknown population. When official license records are used, a large part of the population may be left out of the pool that is being sampled, since in most developing countries many participants in the trading sector are unlicensed. Similarly, simply searching out all traders in a city is a near impossible task, and there is always great uncertainty as to the success of the search. Additionally, statistically minor types or locations of traders or fuel are likely to be either undiscovered or oversampled.

In the backwards-linkage methodology, it is reasonably certain that, given accurate reporting by respondents, all corners of the population have a proportionately equal chance of being sampled. The difficulty lies in assuring accurate responses by interviewees. Careful training of enumerators is required, to ensure that the participant being interviewed discloses all possible supply options, including infrequent ones.

Another benefit of the backwards-linkage methodology is that distances between all nodes of the system are easily measured, and summed over the entire path of the fuel, producing individual aggregate system distances. In the previously used methodology, interrnodal distances are difficult to determine, and even when measured are average system-wide values. In the same way, with this methodology individual price mark-ups along a fuel's path can be measured and spatial variations compared among individual paths, as well as within groups of data points as is usually done with the previously used methodology.

In the same way, the path followed by the fuel from producer to end-user can be easily determined. With the backwards-linkage methodology, the number of links on the market chain is clear, where before this information had to be inferred.

Disadvantages of the Backwards-linkage Methodology

There are a few disadvantages associated with the choice of the backwards-linkage methodology. The greatest of these is the inability of the methodology to provide a way to count the number of participants at each level of the market system. Some useful information is provided by such data, which remains unavailable when using this methodology.

Additionally, the backward-linkage methodology makes no provision for estimating total supplies provided to or from a particular area, except as calculated from woodfuel use data from the woodfuel paths emanating from household users. This reliance on demand-side estimates of supplies as calculated from household consumption data tends to result in an overestimation (Stevenson 1989). Therefore, a 'dragnet' or 'roadblock' survey is generally done in the standard methodology. In these surveys, transporters of woodfuel are stopped when entering a city over a period time and the size of their woodfuel loads determined. Although such surveys generally underestimate supplies (Stevenson 1989), they are useful as a lower limit from which to adjust the overestimate obtained from household surveys. This type of survey could easily be added as a separate component of the backward-linkage methodology.

Finally, the effectiveness of the methodology depends, to an even greater extent than does the standard methodology, on the skill of the enumerators in extracting complete and truthful responses from interviewees. This is not an easy job, and the people used in taking most household and woodfuel surveys in developing countries are not professional enumerators. Therefore, as mentioned earlier, training and monitoring are of particular importance for this methodology.

Following the Path

Using the backwards-linkage methodology, suppliers of woodfuel are weighted by the percentage of the time they are used as the supplier of choice for the buyer being interviewed. Therefore, it is more likely that a supplier used frequently by a buyer will be interviewed than one that is used infrequently. However, there is also a chance proportional to the frequency of use that the lesser-used supplier will be interviewed instead. The structure of the methodology means that some suppliers will be mentioned by more than one buyer. This is particularly true in a neighbourhood, where end users may use one of the few suppliers in the neighbourhood. In these cases, the number of paths is by definition reduced. For example, the two charcoal users interviewed in a Dar es Salaam enumeration area may both happen to use the same supplier for most of their charcoal purchases. In that case, depending on the random weighted selection, one or two different suppliers may be indicated on the fuel path for interviews with seller surveys at the next level of the supply path. Alternatively, one of the end users may get their supplies directly from the producer, or buy their fuel in a different neighbourhood. At times, especially in the smaller towns, traders may supply several neighborhoods. This may occur when the main market is the major source of fuel for area of town, for example.

When two or more paths converge on a single trader, the paths are maintained. In other words, the extra paths are not lost, they just merge temporarily on a particular supplier. When a trader that has been chosen by two or more paths is interviewed, the weighted list of suppliers provided by this trader is sampled once for each path that has identified the seller as its supplier. For example, the weighted list of suppliers of a trader that has been chosen by two paths will be sampled twice. The result of that sampling may fall on two different suppliers, in which case the paths again diverge. Otherwise, one supplier may again be selected by the weighted sampling, in which case the paths remain merged as the next supplier is sampled.

The result of this merging and diverging is that there is not an equal or even predictable number of surveys conducted at each level of the woodfuel path. Overall, the merging of paths means that the number of interviews conducted at the trader level would be less than the number of end users surveyed. To confound things future, there is as often as not more than one trader involved in a woodfuel path – generally one retailer and one wholesaler – so there will be more interviews of sellers than would be otherwise expected, despite the convergence of some paths.

Weighting

The method followed in choosing representative suppliers allows estimates to be made as to the percentage of people the supplier services. This weighting is done using the census-sampling framework. The representative sample of woodfuel users is weighted by the number of households enumerated in the enumeration area. So that path represents that number of households, and the weighting is carried path through the path to the producer. If a supplier has two paths going to him, he is weighted by the households on both of those paths. If the paths from that supplier and another supplier merge later, the next node on the path represents all of the households on the combined paths. In this way, the supply areas can be categorized as to the numbers of households served by each, and ranked in their relative importance to the total supply of woodfuel to the city.

Notes

1. Details about the results of the project can be found in a special edition of *Energy Policy*, Vol. 21, No. 5, 1993.
2. For details about the sample see Maguay, M. and Makbel, M. (1990).
3. Gordon McGranahan of the Stockholm Environment Institute was the source of the idea for and subsequent advice on this methodology.

4 Structure of the Woodfuel Market

Introduction

This chapter examines the structure of woodfuel markets in Tanzania. It tells the human side of the story of the market, and builds a framework on which to place the study of the conduct and performance of the market. It describes the market participants who people the market chain extending from the rural producers and harvesters of fuel, through fuel sellers, and into the urban end users of the fuels. The chapter considers the relations between participants, and relates their organizational and interactive characteristics to the competitive and efficient status of the markets in which they participate. Additionally, the chapter deals with such physical characteristics of the woodfuel system as supply, in particular the location of the supply areas for each city. Although the nature of the market structure chapter is descriptive, such descriptions are necessary for the understanding of the system. Without a good understanding of the human and physical components that underlie the workings of the market system, little can be concluded about the conduct of the participants and the performance of the market.

Demand and Supply

The first step towards understanding the functioning of the woodfuel system in Tanzania is to become familiar with the nature of the supply and demand for woodfuel in the cities being studied. The three cities differ in their woodfuel consumption patterns. Table 4.1 shows the percentage of households using woodfuel in each urban area.[1] The most telling difference among the three urban areas is in the firewood column, where Mbeya is shown to have three to four times more households using firewood than the other cities. Shinyanga has more households using charcoal than the other cities, but the difference is not as acute as in the firewood case. Dar es Salaam has a smaller percentage of users of woodfuel overall, though a slightly higher percentage of firewood users than Shinyanga.

Table 4.1   Household woodfuel use

	1988 population	Households (000)	Charcoal (% using)	Firewood (% using)
Dar es Salaam (n=1600)	1,241,000	314.3	74.6	16.8
Mbeya (n=620)	132,000	33.7	78.9	58.5
Shinyanga (n=450)	47,000	20.3	85.1	13.8
Average of all cities	–	–	77.4	26.0

Figures will not add to 100% because many households use both fuels.

Source: Hosier and Kipondya 1993.

Scarcity and Consumption Patterns

In contrast to the other two cities, Shinyanga has very little in the way of woodfuel resources in the immediate area. According to forest officers, the area surrounding the city has been deforested for at least 50 years and there are no local plantations that can be harvested for firewood. However, there is a small plantation of local *Acacia spp.* under forest department auspices not far away that is being reserved and, if possible, added to. Consequently, most of the woodfuel that is sold to end users in Shinyanga is imported into the city from outside the surrounding area. Though, as shown in Table 4.2, most of the firewood that is used by households in the town is collected or cut by the user, usually small pieces from shrubs on their farms, the amount is relatively small. Most of the firewood sold in Shinyanga goes to informal sector industries (especially beer brewing).

In Dar es Salaam, not surprisingly, somewhat fewer households use woodfuel. In particular, the use of firewood is limited, and 40 per cent of that used is collected or cut by the user from a tree at their home or farm or from discarded packing cases and other industrial leavings. Charcoal use in Dar es Salaam is also slightly lower than in the other municipalities surveyed. Lower overall woodfuel use there is due to greater access to and use of 'modern' fuels such as electricity, kerosene, and LPG. The subsidized prices of these fuels, in particular electricity and LPG, are lower than the effective price of charcoal or firewood, encouraging their use (Hosier and Kipondya 1993).

Table 4.2 Percentage of fuel by weight obtained by households from each source

	Dar es Salaam Charcoal	Dar es Salaam Firewood	Mbeya Charcoal	Mbeya Firewood	Shinyanga Charcoal	Shinyanga Firewood
Seller in town	86.7	59.1	87.8	49.5	99.0	21.4
Roadside seller	5.2	0.1	9.8	1.5	0.0	0.0
Rural producer	8.1	0.7	2.4	6.5	1.0	0.0
Cut/gathered by user	–	40.1	–	42.5	–	78.6

The differing consumption patterns of each city require different supply patterns as well, which will be explored in the next section.

Supply Areas

Figure 4.1 shows the areas supplying each of the cities with woodfuel. Table 4.3 describes the amounts supplied from each area, and gives the average distance from a producer at his harvest point in that area to an end user in the city. Though most areas supply both firewood and charcoal, firewood from the parts nearer to town and charcoal from the further areas, generally each area 'specializes' in one or the other fuel.

Dar es Salaam

Dar es Salaam has the most supply areas, which were categorized by the road transporters must use to get to them. The area to the west of Dar es Salaam, along the Morogoro Rd through Chalinze and north up the Tanga Rd provides the largest part of the charcoal supply for the city. The fuel from this area also travels the furthest distances of any fuel supplied to Dar es Salaam. The area provides a small amount of firewood, which also travels a greater than average distance. The clear reason for this is that the road serving this production area is by far the best road of any area, allowing easy access to producers.

The Kilwa Rd/Kibiti area, due south of Dar es Salaam along a partially tarred road, and the Kisarawe/Chanika area to the southwest of the city along a graded dirt road are the other areas which each provide about a quarter of the city's needs. These areas are wetter and more lush than the areas to the west of the city, and have long been providing woodfuels to the city. Almost every man in almost every village in these areas sells woodfuel for

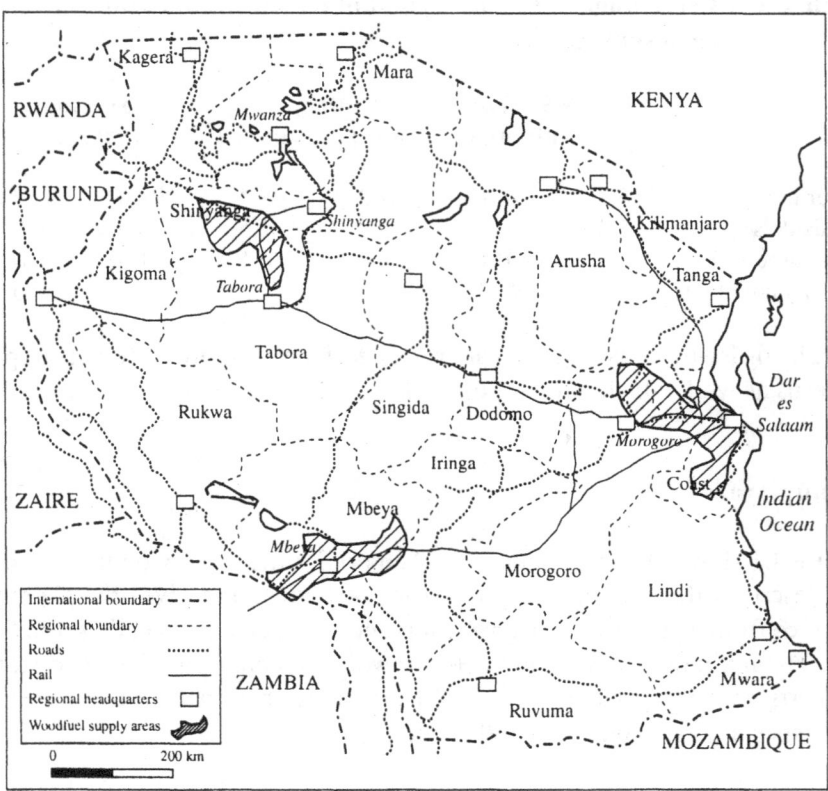

Figure 4.1 Approximate woodfuel supply areas: Dar es Salaam, Mbeya and Shinyanga

supplementary income at times, or so it is supposed by some of the producers. Together they supply almost 90 per cent of the firewood brought into town, often from areas quite distant and along poor roads. The distances travelled by charcoal and firewood from the Kisarawe area are similar, though some villages seem to specialize in one fuel or the other.

The Bagamoyo Road area provides little fuel, though the distances are somewhat shorter and the road not too bad. The area appears to have been an early supplier of woodfuels to Dar es Salaam, but the deterioration of the road and reduction of traffic to Bagamoyo may have limited the supplies from this area. The area is less lush than the areas to the south and west, and less forested.

Table 4.3 Fuel supply areas – percent of supply reaching end users

| | Percent of supply | | Average distance to end user (km) | | |
	Charcoal	Firewood	Charcoal Mean (SD) n		Firewood Mean (SD) n	
Dar es Salaam	–	–	102.4 (48.9)	67	69.2 (38.7)	29
Bagamoyo Rd Area	0.0	3.0	–		44.7 (3.3)	2
Chalinze/Moro-goro Rd area	43.6	7.0	130.8 (47.6)	26	80.0 (0.0)	1
Kilwa Rd/Kibiti area	27.0	61.2	95.4 (41.4)	4	68.4 (47.1)	17
Kisarawe/Chanika area	25.8	27.7	64.5 (33.7)	15	70.7 (25.0)	9
Kimbiji area	3.4	1.17	3.5 (19.1)	2	–	
Upcountry	0.2	0.0	–		–	
Mbeya	–	–	116.0 (36.7)	24	20.0 (38.0)	22
Chunya District area	15.1	0.1	100.5 (34.4)	12	–	
Mbozi District area	27.6	0.7	119.7 (30.8)	7	–	
Usangu area	50.1	10.8	145.3 (35.4)	5	90.3 (76.5)	3
Mbeya Municipal area	7.2	88.4	–		8.0 (7.6)	19
Shinyanga	–	–	172.6 (31.6)	26	105.0 (56.7)	7
Kahama District area	88.6	0.0	176.8 (13.1)	19	–	

Mbeya

Mbeya gets half of its charcoal and nearly all of the firewood that is brought in from out of town from the Usangu area, east of the city. It is accessible partly by the main road, but the roads become worse quickly upon leaving the tarred road. The distances to this area are the greatest of any for charcoal supply to Mbeya, but either because of the partial access by good road, or because the area has been providing fuel for a long time, with the distances getting longer over the years, the area is popular. The Mbozi and Chunya areas provide the bulk of the rest of the charcoal supply for Mbeya, with the shortest distances but the worst roads belonging to Chunya, to the northwest of the city, and slightly better roads to Mbozi, west of the city towards the Zambian border. There seems to be little activity to the south of the city, on the good road linking Tanzania and Malawi. This area has long supported intensive agriculture and cash crop plantations, and there are no indigenous trees to be seen from the road.

Shinyanga

Shinyanga is notable mostly for the significantly longer distances travelled by the fuel used. Its fuel travels at least 50 per cent further than that going to the other cities. Shinyanga's charcoal needs are served primarily by fuel from the Kahama district, southwest of the city. Some charcoal also comes in from Nzega District in Tabora region, also southwest of town. Both are accessed by graded dirt roads, and some parts of the Nzega area are also served by rail. Most of the firewood brought into Shinyanga enters by rail, and this efficient transport mode allows the greater distances to be viable.

Distances Travelled by the Fuel

These distances, in an area widely considered to be deforested are less than those reported by the 1986 World Bank survey for firewood going to the deforested areas of Mwanza and Arusha (i.e., 200 to 300 km by road, and for charcoal up to 500 km by rail) (Hosier et al. 1990, Luhanga and Kjellstrom 1988, Openshaw 1984, World Bank 1984, Kaale 1983). They are also less than the 500 km limit travelled by charcoal to Dakar, Senegal, and comparable to the 170 km covered by firewood going to Bamako, Mali (Bertrand 1985). These comparisons suggest that there is the possibility of expansion of the service area for woodfuel going to Shinyanga and, perhaps, the other urban areas, particularly given access to rail transport and improved roads. The desirability of this expansion, and whether to encourage, discourage or ignore it from a policy standpoint is addressed in a subsequent section.

The 1986 survey reported woodfuel transport distances to Dar es Salaam ranging from 70 to 200 km, with an average of 100 to 150 km (World Bank 1987b). These figures accord well with the average of 102 km for charcoal and 69 km for firewood reported by the 1990 study, though the 1986 results are somewhat higher on average.

Table 4.4 and Figure 4.2 show the average distance from a producer at his harvest point to an end user in the city. Charcoal travels over twice as far as firewood on the average, and fuels going to Shinyanga travel one and a half times as far as they do to other cities. Though there is much variation in the

Table 4.4 Distances travelled by fuels

	Total distance travelled (km)	Trader to end user (km)			Harvest area to main road (km)		
	mean	mean	(s)	n	mean	(s)	n
Charcoal	120.6	0.57	(0.76)	194	6.61	(7.96)	89
Dar es Salaam	102.4	0.35	(0.52)	90	5.52	(4.17)	46
Mbeya	116.0	0.56	(0.74)	62	7.04	(12.1)	19
Shinyanga	172.6	1.03	(1.0)	42	8.35	(9.4)	24
Firewood	54.8	1.28	(4.0)	86	8.30	(7.94)	40
Dar es Salaam	69.2	1.27	(5.7)	41	9.93	(8.62)	28
Mbeya	20.0	0.81	(0.88)	37	2.99	(2.29)	10
Shinyanga	105.0	3.44	(2.0)	8	12.0	(4.24)	2

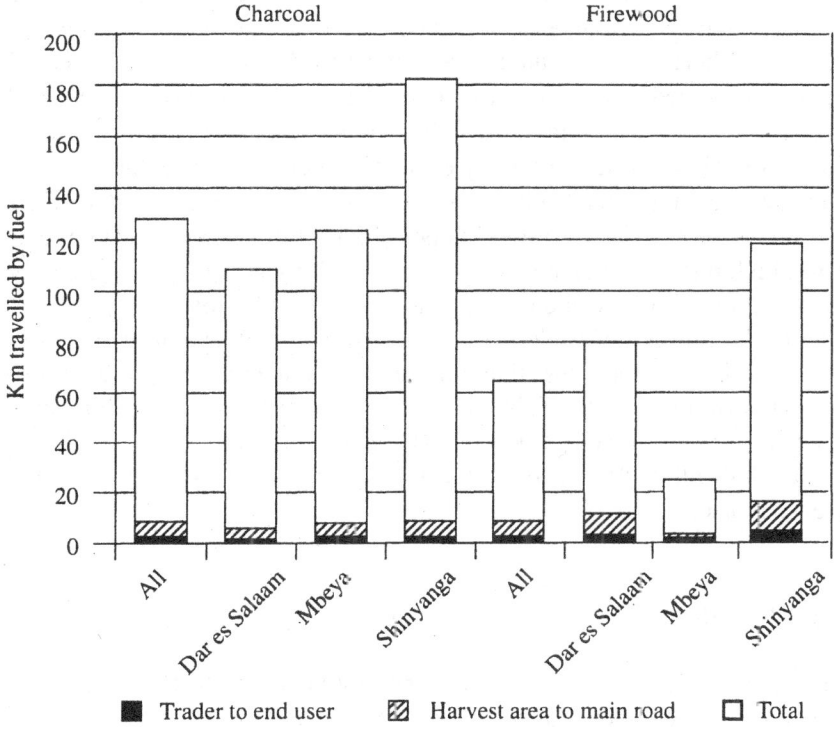

Figure 4.2 Distance travelled by fuel

results, it is interesting to note that in all cases the distance from the harvest point to the road is 5 per cent of the total distance travelled for charcoal and 15 per cent for firewood. The road quality of these access roads is universally poor, and perhaps there is a tolerance limit for transporters or traders.

Seller Frequency

Another interesting aspect of Table 4.4 is that the reported distances between trader and end user bear out the observations of size and frequency of traders in the different cities.

End users have the shortest trip to charcoal traders in Dar es Salaam, followed by those in Mbeya. In Shinyanga, the distance travelled is greater, double that of Mbeya and three times that of Dar es Salaam, and there are fewer sellers. This sparks a question as to the reason why purchasers in Shinyanga spend more time in the purchase of their fuel than users in the other cities. If the value of time is the same in each of the towns, the expectation would be that there would be an increase in the frequency of sellers to equal the distance travelled by users in those towns. This anomaly may be explained by the lower density of Shinyanga, where more traders would reduce the number of user households served by each seller below an acceptable profit-making level. It might also indicate a restriction in the ease of entry at the retail level of the market system. This possibility will be explored in a later section. The longer distance between traders in Shinyanga may also actually be explained by a lower value put on time in that smaller town.

Due to the number of traders, firewood sellers in Mbeya are situated in proximity to their patrons, less than a kilometre on average, and a bit fewer and farther between in Dar es Salaam (over one kilometre distant). Shinyanga again has the greatest distance between firewood users and sellers, at almost three and a half kilometres, primarily because few households buy firewood for everyday use.

Transport Modes

The transport of firewood is done primarily by headload in Shinyanga and Mbeya, since many traders collect and cut their own fuel in these cities. Fuel is brought in from distant areas by truck in Dar es Salaam and Mbeya, and by truck and train in Shinyanga. Figure 4.3 describes the distribution of transport

Structure of the Woodfuel Market 61

modes for firewood, and Figures 4.4 and 4.5 show transport modes used by wholesalers and retailers of charcoal. Generally, when transported by truck, it is collected at the harvest site by the wholesaler and transported to the city. Other times, the harvester transports the fuel to the road- or rail-side himself, where it meets a city-bound truck or train and is transported into the city.

Charcoal is moved almost exclusively by truck. In Dar es Salaam and Shinyanga charcoal is also transported by train. However, perhaps because it is generally done illegally by loading and unloading outside of official stations and paying only the necessary rail workers, the survey did not identify any users of that mode of transportation. The pushcart, wheelbarrow and headload percentages for retailers reflect intricate transport in all cities. Again, the norm is for the fuel to be picked up by a wholesaler at the kiln site, and transported directly to the city. However, especially on the roads leading to Dar es Salaam, the incidence of 'roadside sellers', is significant. In these cases, the fuel is transported to the roadside by the producer or a wholesaler, who then sells it to another wholesaler or transporter for the final trip into town.

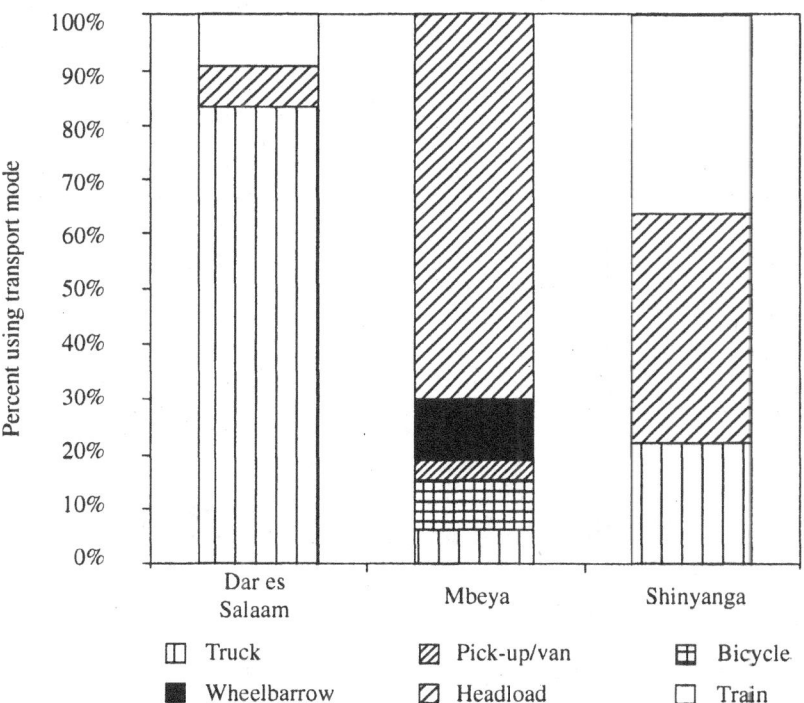

Figure 4.3 Transport modes for firewood

62 *Woodfuel Markets in Developing Countries*

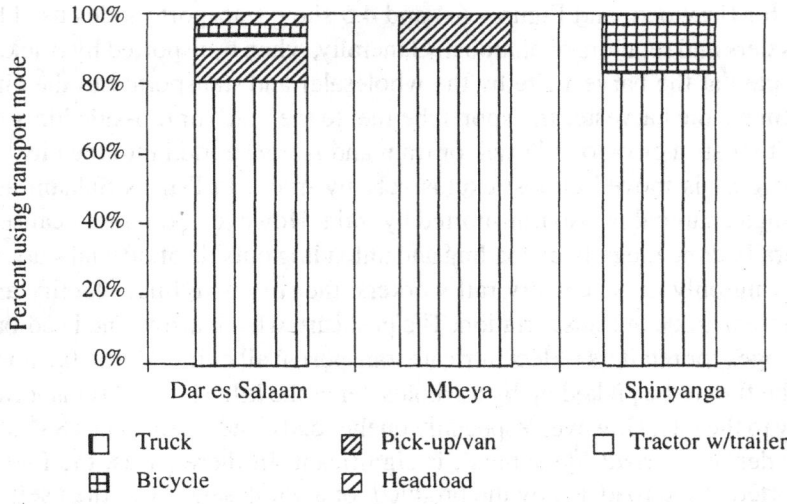

Figure 4.4 Transport modes for charcoal wholesalers

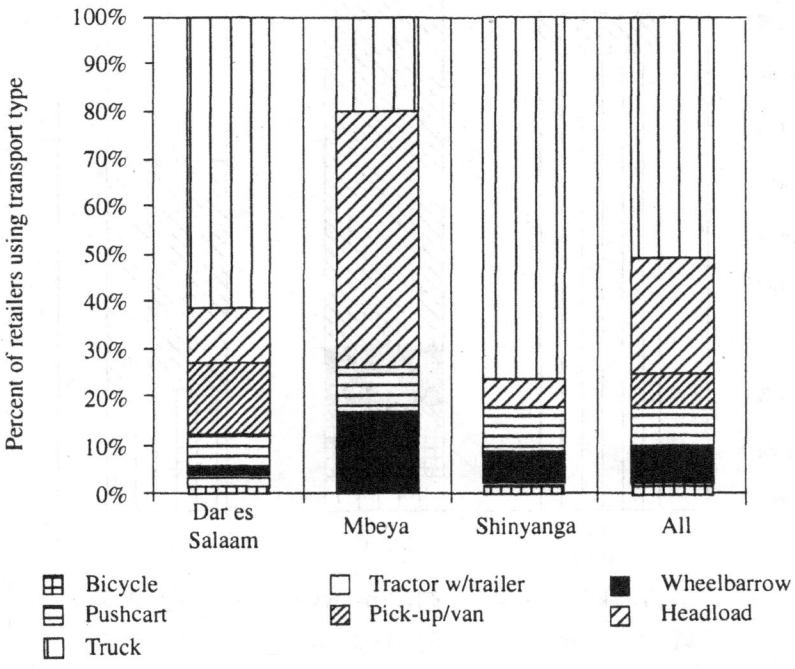

Figure 4.5 Transport modes for charcoal retailers

The difference between the transport modes of wholesalers and retailers of charcoal, and the differences between the cities, are easy to see. The most obvious differences are the heavy use of non-mechanized personal ('head-loads') and wheelbarrow transport by Mbeya retailers, and the use of bicycles by Shinyanga wholesalers. In both cities, there are transport bottlenecks caused by the lack of trucks in the country, and competition with other uses.

Transporters' trucks were generally purchased second-hand 10 or more years ago, and are at least 20 years old. The trucks are usually 7- or 10-ton, and carry 150 to 200 bags of charcoal or approximately 200 pieces of firewood per trip. A few transporters said that they carried woodfuel because their vehicle was unsuited to any other cargo, because of size or condition. Certainly, the 'woodfuel fleet' is a rusty one, and from all appearances held together by wire and luck. However, although the more dependable and higher value vehicles move other sorts of cargo, there does not seem to finally be any problem in supplying the traders via truck. In one remote supply area visited twice, several months apart, by the survey team, the same vehicle was encountered both times broken down along the road. However, in both cases the truck was eventually repaired and completed its delivery.

Supply System

Length of Market Chain

The supply system for woodfuels in Tanzania is relatively straightforward. The average path linking a woodfuel end user to the harvester of the tree from which the fuel originated involves three or four participants (including the end user). In the case of firewood the path can be as short as two nodes, though in some charcoal cases it is as long as five nodes. A typical path consists of a harvester/charcoal maker, a trader, perhaps a retailer, and an end user.

As seen in Figure 4.6, about half of charcoal end users in Dar es Salaam are served by a path that includes at least a second middleman, usually a small-scale retailer operating in a market or selling from the home. In Mbeya, three-quarters of end users are on a path that includes four or five nodes, and in Shinyanga about a third. Mbeya has slightly longer chains for both fuels. For charcoal it hinges on the great number of end users who are served by a market chain which includes four nodes (two sellers in-between the producer and end user). In Shinyanga, far more end users buy directly from a trader who has bought from a producer himself.

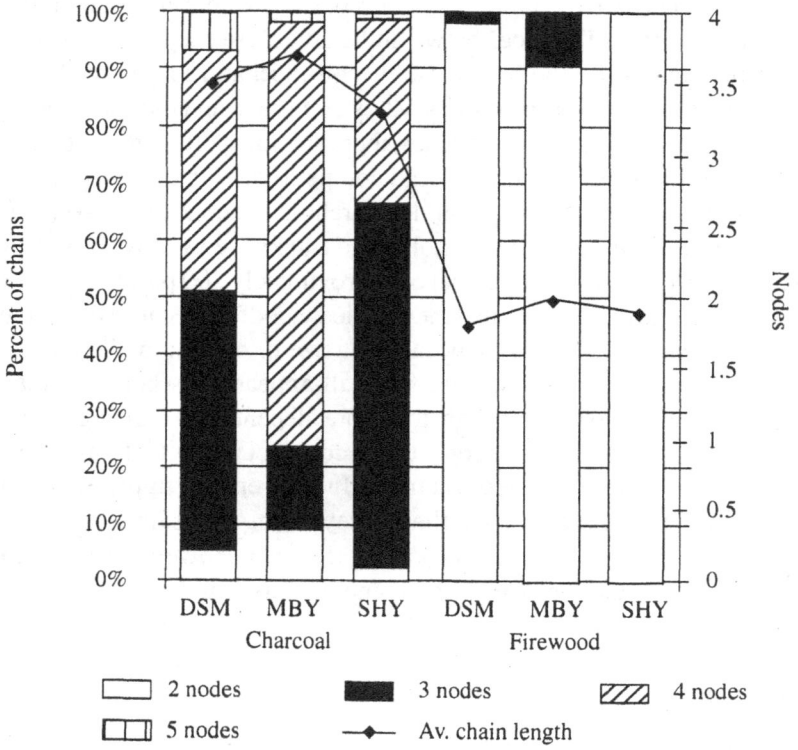

Figure 4.6 Supply of woodfuels – number of nodes in supply system for each city

The average number of nodes in firewood supply chains is essentially the same in the three cities, though the distribution is slightly different in each. In particular, Mbeya has fewer end users collecting their own fuel, so more of them buy from a trader. For the most part, the trader hires transport, even though transporters act as wholesalers or even retailers in slightly over 10 per cent of the cases. These percentages reflect the organizational differences that exist in the woodfuel trade of each town, such as the existence of a large wholesaler in Mbeya that provides nearly 20 per cent of charcoal users with their fuel directly or through retailers. It also reflects the variation in the integration of the systems of the different urban areas. These issues will be discussed in some detail in a subsequent section.

The supply system is similar to those identified in Haiti (Stevenson 1989), Senegal (Ribot 1988) and other Sahelian countries (Bertrand 1986, Diallo and Fell 1983) with variations particular to the circumstances of the particular country. It is also similar in length and composition to those serving urban areas in food provision (cf. Jones 1972, Southworth et al. 1979, Schmidt 1982 and others).

Participant Profiles

The social and economic profile of players from all stages of the system is displayed in Tables 4.5 and 4.6. A profile of a typical participant is a middle-aged married male with a primary school education who began in the woodfuel business in the last five years, and derives his primary income from his woodfuel business or, in the case of producers, farming. Before becoming a woodfuel supplier he most likely had another small-scale business or farmed, and he entered the woodfuel business because he saw it as a profitable alternative to what he was doing.

Of course, there are many exceptions to this profile. One of the most significant is the large of numbers of women participating in the woodfuel supply system in Mbeya as traders and harvesters. This may be due to the high firewood use in that city, and to the plantations around and within the city that provide the great majority of the firewood supply. The women are able to walk to the plantations, cut or gather branches and small trees themselves (or with their children), and carry the fuel back to town to sell in the market. The costs and infrastructure involved are minimal, and therefore allow the women to participate more fully than in the other areas, where they are generally charcoal retailers on a small scale, or sell branches from their own trees. The participation of women is also reflected by the high number of sellers who have gone from being housewives or mothers to woodfuel vendors, and who are in that business particularly because it allows them to concurrently watch their children and do other household tasks, as seen in Table 4.6.

Another, related, exception is the lower age and years of experience of traders in Mbeya, and the higher age of participants in Dar es Salaam. All of the youngest traders of firewood in Mbeya are female, and their experience is consequently briefer. In Dar es Salaam, the average age of traders is higher, with many fewer players under 30 years of age than in either Mbeya or Shinyanga. Along with the slightly greater length of experience of Dar es Salaam traders, this seems to imply that the woodfuel trade may be more established there. This is upheld by the correspondingly long years of experience of the Dar es Salaam producers.

Table 4.5 Participant profiles

	Sellers			Producers		
	DSM n=157	MBY n=103	SHY n=50	DSM n=98	MBY n=47	SHY n=34
Sex (%)						
Male	82.2	23.3	68.0	98.0	70.2	97.1
Female	17.8	76.7	32.0	2.0	29.8	2.9
Age						
Mean (%)	36.8	32.6	33.6	37.1	35.3	34.3
	(9.3)	(9.7)	(8.2)	(12.4)	(13.5)	(9.73)
<20	0.0	5.0	0.0	5.3	2.3	0.0
20–29	22.4	33.7	34.7	24.5	45.5	36.4
30–60	75.0	59.4	65.3	66.0	47.7	63.6
>60	2.0	2.0	0.0	4.3	4.5	0.0
Years in business						
Mean (%)	3.0	2.4	2.6	11.3	3.1	2.3
	(3.2)	(2.4)	(2.3)	(9.8)	(3.6)	(2.3)
<1 yr	22.9	36.9	34.0	2.0	34.0	23.5
<5 yr	82.7	89.4	84.0	34.4	78.8	88.3
5–10 yr	15.9	9.7	16.0	24.2	12.8	8.8
>10 yr	1.3	1.0	0.0	41.4	8.5	2.9

	Transporters	
	DSM n=20	MBY n=6
Sex (%)		
Male	95.0	100.0
Female	5.0	0.0
Years in business	36.9	46.7
	(6.8)	(5.0)
Mean (%)	4.3	9.2
	(1.2)	(5.2)
<1 yr	0.0	0.0
<5 yr	55.0	33.4
5–10 yr	35.0	16.7
>10 yr	0.0	50.0

Structure of the Woodfuel Market 67

Table 4.6 Employment history, reasons for being in woodfuel business

	Sellers			Producers		
	DSM[a]	MBY[b]	SHY[c]	DSM	MBY	SHY
Primary income earning activity						
Woodfuel seller/transporter/harvester	46.5	59.2	58.0	22.2	42.6	32.4
Farmer/ agricultural worker	24.8	29.1	22.0	76.6	55.3	67.6
Government worker	10.8	1.9	10.0	–	–	–
Own business	17.8	9.7	10.0	–	–	–
Other employment	–	–	–	1.0	2.1	0.0
Previous occupation						
Student	3.2	6.8	12.5	–	–	–
Farmer/ agricultural worker	33.5	27.3	33.3	–	–	–
Housewife/ mother	11.0	41.7	12.5	–	–	–
Casual worker	12.3	2.9	4.2	–	–	–
Government worker	3.2	1.0	4.2	–	–	–
Other employment	2.6	0.0	8.3	–	–	–
Own business	34.2	20.4	25.0	–	–	–
Reason is in woodfuel business						
Supplementary income	38.3	27.7	32.0	91.9	59.6	88.2
Family business	–	–	–	7.1	8.5	8.8
Best source of primary income	34.5	24.7	48.0	1.0	31.9	2.9
Can watch children/do other economic activity	16.8	44.6	16.0	–	–	–
Retired	10.4	3.0	4.0	–	–	–

Notes: a = Dar es Salaam; b = Mbeya; c = Shinyanga.

Producers

Relationship of producers to land Other differences exist between producers in the three cities. The most telling difference concerns their relationships with the wholesalers and to the land, from which they cut and produce their product. As reflected in Table 4.7 and Figure 4.6, in Dar es Salaam producers are almost exclusively local farmers who produce firewood and charcoal during the dry season as an additional source of income. The resultant stability of residence of producers helps to support the evidence of the longer extent of market involvement of producers there.

In contrast, as Figure 4.6 graphically shows, in Shinyanga and Mbeya there are more producers who are professionals, in that they produce woodfuel as their primary income source. They are also more likely to be from somewhere other than the region in which they are working. In addition, these producers are more likely to be working under the license of a wholesaler or, in the case of Shinyanga, a family or village license. This means that their relationship with the land and with the local inhabitants can be very different from producers in Dar es Salaam. Table 4.7 describes producers in more detail. Dar es Salaam producers are seen to be far more likely to be local residents than producers in the other cities, and those in Mbeya are the least likely to be from the local area. This corroborates the evidence of the prevalence of Mbeya wholesalers to control harvesting licenses, and supports the idea that most harvesters there are professional producers unconnected to the land or harvest area.

Table 4.7   Primary income source of woodfuel producers

	Dar es Salaam	Mbeya		Shinyanga
		male	female	
Primary economic activity (%)				
Charcoal/firewood supplier				
Charcoal	14.9	54.2	0.0	33.3
Firewood	37.5	0.0	53.8	28.6
Farmer/agricultural worker				
Charcoal	83.6	45.9	100.0	66.7
Firewood	62.5	88.9	46.2	71.4
Other employment				
Charcoal	1.5	0.0	0.0	0.0
Firewood	0.0	11.1	0.0	0.0
Local resident (%)	91.0	44.0		70.4

Experience In Shinyanga and Mbeya, there are many producers who have been producing (or harvesting – the terms are used interchangeably here) for less than a year, and a great majority of them have been producing for less than five years. This could reflect a new emphasis on woodfuel supply in those areas, due to a new awareness of its profitability or an increase in demand. It also might reflect the quick movement of the harvest areas so that new participants are involved every few years as one supply area is depleted and another brought into production. Since most producers in Shinyanga are primarily farmers, they are not in general likely to move with the production. Alternatively, the short time period may just indicate quick turnover, as participants tire of the work or find easier or more lucrative employment.

In the case of Mbeya, however, there are more harvesters who produce for their primary income, and so they are more likely to follow the supply areas as the areas move due to seasonality or depletion. In fact, in conversations with charcoal burners in the Mbeya area during the dry season, many were found to live on site, alone or with other burners, and to move to other locations in the rainy season where there was less rain and better roads. Others mentioned that they sometimes went to Zambia to produce charcoal, the political situation permitting. In Shinyanga, many charcoal producers also live on site, sometimes with their families. They burn several kilns, and then move on, or sometimes farm the areas that have been cleared from their charcoal making. Only in Dar es Salaam were all charcoal producers found to 'commute' to their kilns from their permanent residences.

In Dar es Salaam, the suppliers of firewood have been in the business even longer than the charcoal burners. This longevity, however, is dependent on the area in which the supplier works and lives. 4.8 displays the length of time a supplier has been in business by supply area for each city. Though the numbers are not particularly robust, they indicate that there is some variation between areas and, to a lesser extent, between fuels. In Dar es Salaam in particular, there are some areas which have producers who have been in business for a particularly long time, such as charcoal burners in the Kilwa Rd/Kibiti area and the Kisarawe/Chanika area. This makes sense in that these are the areas that have long provided Dar es Salaam with its fuel, and include the outer parts of some Dar es Salaam districts.

In Mbeya, the Usangu area charcoal producers appear to have been in the business longer than their counterparts in other areas. Since Usangu is also the area that provides the most charcoal to Mbeya, this might indicate that this area was the first to begin the supplying of the fuel. There are no real differences between producers in the Shinyanga area. The supply areas are

70 *Woodfuel Markets in Developing Countries*

discussed further, and a map of them is provided, in the section on supply.

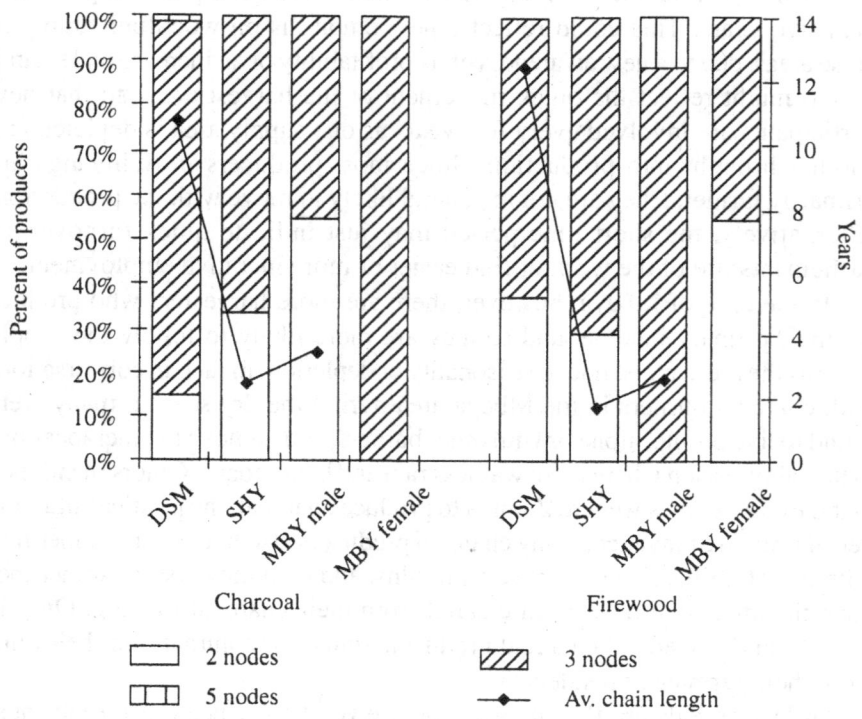

Figure 4.7 Primary occupation and longevity of fuelwood producers

Transporters

Finally, transporters are not well represented in this set of surveys because only those who acted as a middleman for the fuel in a path were interviewed. Since it turned out that only slightly over 11 per cent of households and about 10 per cent of traders are supplied directly by transporters (about 1 per cent of which are petroleum tankers back hauling from upcountry), not many were indicated on the paths followed. Most transport is hired by the traders, and, though some of those transporters do much of their business transporting woodfuel, they are seen not so much as players in the system as functionaries. For that reason, though the function of transport will be discussed in this

Table 4.8 Differences in years in business by fuel supply area

	Charcoal mean (SD) n		Firewood mean (SD) n	
Dar es Salaam				
Bagamoyo Rd area	–		32.3 (2.1)	2
Chalinze/Morogoro Rd area	8.0 (7.6)	26	15.8 (0.0)	1
Kilwa Rd/Kibiti area	13.2 (10.3)	24	11.9 (9.3)	20
Kisarawe/Chanika area	12.5 (12.1)	15	8.6 (6.6)	9
Kimbiji area	5.8 (0.1)	2	–	
Mbeya				
Chunya District area	3.1 (3.4)	12	–	
Mbozi District area	2.5 (2.8)	8	–	
Usangu area	6.0 (4.4)	8	1.7 (2.7)	3
Mbeya Municipal area	–		2.9 (3.8)	19
Shinyanga				
Kahama District area	2.2 (1.9)	19	–	
Nzega District area	3.5 (3.6)	8	1.9 (1.3)	6
Shinyanga Municipal area	–		0.8 (0.0)	1

study, the details of transporters as participants in the system will not. Nonetheless, some details can be determined about those who are players.

Transporters tend to be a bit older than other participants in the supply system. They tend to have been in the transport business for a long time, whether carrying woodfuel or other cargo. Many describe their woodfuel transport business as a family business, and most depend on woodfuel transport for their primary income. Many were drivers or transporters before they began carrying woodfuel, but about half were farmers or businessmen. Most find the business profitable, and a good way of using their vehicle.

Size Distribution

The relative size distribution of competitors is an important characteristic of market structure. Table 4.9 describes the amount of fuel bought and sold at each stage of the supply path. Total amounts sold to household end users are calculated from household survey figures, obtained by weighing fuel on two consecutive days to determine the amount consumed. The average quantity

Table 4.9 Kilograms of fuel bought and sold yearly at each stage of the supply system

	Dar es Salaam (thousand kg)		Mbeya (thousand kg)		Shinyanga (thousand kg)	
	Charcoal	Firewood	Charcoal	Firewood	Charcoal	Firewood
End user (yearly)						
Kg consumed per household	1.2	1.7	0.97	1.8	0.96	0.51
(household size)	(4.3)	(4.3)	(4.5)	(4.5)	(4.9)	(4.9)
Kg bought per capita	0.28	0.40	0.22	0.40	0.20	0.10
Total household purchases (000)	261	125	16.3	46.3	4.33	2.33
Trader (yearly)						
Kg bought per trader	123	169	80.6	25.6	147	159
Kg sold to households/trader	85.1	119	69.8	23.1	110	1.67
Secondary trader						
Kg sold per retailer	67.0	–	28.1	–	112	–
Kg sold per wholesaler/retailer	143	–	540	–	144	–
% selling some fuel to retailers	55.3	34.9	19.4	8.3	43.9	44.4
% buying some fuel from wholesaler	30.7	9.3	77.3	8.1	24.4	*
Producer/harvester (yearly)						
Kg sold per producer	60.0	169	42.1	23.9	36.3	13.9

* too few firewood users were encountered to determine this value.

of fuel bought by households varies by city, with Dar es Salaam residents consuming more charcoal per capita and per household than those in the other cities. Shinyanga households consume less of each fuel than those in Dar es Salaam and Mbeya, perhaps because of the scarcity of fuel in the area. There is some error in these numbers due to seasonal differences in fuel choice and, to a lesser extent, in usage rates. Firewood-using households tend to switch to charcoal during the wet season, and the values shown reflect dry season levels of use only.

Shinyanga

In both the firewood and charcoal markets, traders in Shinyanga buy and sell more per year than their urban counterparts elsewhere, but in smaller bunches. Therefore Shinyanga traders entertain more customers per trader. Numerous very small-scale traders in markets and homes are a frequent sight in Dar es Salaam and Mbeya (for example, one residential area in Dar es Salaam of about 100 households contained 10 small-scale traders). The norm in Shinyanga is a relatively large trader located by the railway station in the centre of town or in a cleared area in the residential sections.

Mbeya

Conversely, Mbeya is notable for its low volume retailers. This is particularly the case for firewood traders. As mentioned in the previous section, female traders who collect or cut the wood themselves and sell it in the local markets are the most frequently encountered firewood sellers. There are many participants, and many only sell three or four small bundles (one headload) in a day. This contrasts with Shinyanga and with Dar es Salaam, where the charcoal retailers are often very small-scale, but firewood traders are more infrequent and larger-scale, especially when sales to the informal sector are included. In all cases but Shinyanga, numerous small-scale traders are prevalent for the fuel preferred by households, and fewer large-scale traders are in evidence for the less-preferred fuel.

Secondary Traders

It is interesting to note the prevalence and influence of secondary traders (wholesalers or wholesaler/retailers) in each urban area. In Dar es Salaam, about half of the charcoal traders also (or, in rare cases, exclusively) sell to retailers, and about 30 per cent of the traders buy some or all of their fuel from other wholesalers. The average sales for a retailer who does not do any wholesale business are about half that of a wholesaler, which takes into account the smaller units many retailers sell in. In Shinyanga the wholesalers only sell an average of about 22 per cent more than retailers, emphasizing the point that traders are relatively large, in general. The average charcoal trader in Shinyanga has about 40 bags of charcoal in stock, as compared to 17 in Mbeya and 32 in Dar es Salaam.

Mbeya cooperative In Mbeya, there is a huge disparity between the average sales of retailers and wholesalers. This is due primarily to one large wholesaler, a 'cooperative' that accounts for more than a quarter of the charcoal sold in the city. This organization is a loose consortium of several businessmen who run a completely integrated charcoal business, from production to sales. The members of the group are each responsible for their own part of the business that determines their profits. However, they cooperate by sharing equipment, selling space, official licenses and fees, and by helping each other out when needed. The members of the group organize producers in the rural areas, provide them with tools and food (later deducted from their earnings), a license to work under, sacks, and guaranteed purchase (they also require a minimum output). The producers are paid approximately the farm gate market rate for their production. The group members arrange for transport and go with the transporter to pick up and pay for the charcoal. They then hire a trader for the joint depot centrally located in Mbeya, and share supervision of sales.

Additional evidence is displayed in Table 4.10. It shows the size distribution of traders based on their annual gross turnover. The deviations from average sales in all cases are large, but they are particularly so in Mbeya. There is a great variation between sellers, with the more important wholesalers' sales being relatively over represented. For charcoal, the distribution of wholesalers varies between cities. In Dar es Salaam, the distribution is nearly normal, while in Mbeya and Shinyanga it is skewed, indicating the existence

Table 4.10 Distribution of sellers by gross annual turnover

	Wholesalers			Retailers		
	DSM	MBY	SHY	DSM	MBY	SHY
Charcoal						
Mean (10^6 Tsh)	3.6	3.1	3.2	2.6	8.3	2.9
Median (10^6 Tsh)	3.6	2.7	2.3	2.4	8.8	2.5
SD (10^6 Tsh)	1.5	2.7	2.1	1.3	7.3	2.0
Skewness	0.04	1.02	2.07	1.07	-0.10	1.09
Kurtosis	-0.12	1.32	4.38	2.09	-2.95	1.47
Firewood						
Mean (10^6 Tsh)	0.66	0.20	0.99			
Median (10^6 Tsh)	0.35	0.13	1.1			
SD (10^6 Tsh)	0.79	0.23	0.52			
Skewness	1.78	3.04	-0.77			
Kurtosis	2.38	11.6	0.42			

of a few important wholesalers with a relatively high market share. Charcoal retailers in Mbeya demonstrate a normal distribution, while those in Dar es Salaam and Shinyanga demonstrate skewed distributions.

Firewood Retailers

In the case of firewood, sellers in Mbeya are highly skewed to the right. This is a result of the existence of the numerous small-scale retailers already discussed. If the wholesalers could be separated from the retailers here, the skew would be decreased greatly. In Dar es Salaam, there is a skewed distribution, for the same reasons, as the neighbourhoods of Dar es Salaam are full of small retailers. In Shinyanga, the absence of such retailers is confirmed in the leftwardly skewed distribution of sellers.

Size of Charcoal Sellers

Table 4.11 gives the percentage of different sized charcoal sellers to be found in each of the cities. Wholesalers in Dar es Salaam are likely to be large or medium sized, while wholesalers in the smaller cities of Shinyanga and Mbeya are most likely to be of medium size. In Shinyanga the size of wholesaler categorized as medium for the comparison of the three cities is actually large for the relative size of that city. Retailers tend to be small in all three cities, but particularly so in Mbeya. In contrast, Dar es Salaam and Shinyanga have a higher percentage of medium-sized sellers.

Table 4.11 Size distribution of charcoal sellers (by number of bags sold)

	Seller scale (% of total by number of bags sold)		
	Small (<200/mo)	Medium (200-450/mo)	Large (>450/mo)
Dar es Salaam			
Wholesalers	14.3	34.1	51.6
Retailers	49.3	32.3	18.4
Mbeya			
Wholesalers	22.8	56.9	20.3
Retailers	69.1	13.0	17.8
Shinyanga			
Wholesalers	20.4	75.8	3.8
Retailers	50.4	39.3	10.3

In various degrees, charcoal sellers in all three cities characterize one of their biggest problems as being a plethora of other sellers. Almost three-quarters of retailers in Dar es Salaam see this as a problem, compared to about 40 per cent of retailers in the other two towns. This reflects the earlier observation of the existence of many retailers there, but does not guarantee competition and particularly efficiency, as there seems to be an oversupply. Approximately 60 per cent of wholesalers in Dar es Salaam and Mbeya also saw the number of wholesalers as a problem, but fewer than 20 per cent of Shinyanga wholesalers agreed. This is interesting in light of the fact that the traders' margins (detailed in chapter 6) in Dar es Salaam and Mbeya are much higher than those in Shinyanga.

Firewood Wholesalers

In all three cities, there seems to be small number of firewood wholesalers that sell to a small number of retailers, but most traders get their own fuel, in or out of town, and sell to end users or a smaller retailer. The properties of firewood allow people to be their own suppliers, on a small scale. Again, sellers in Dar es Salaam saw the number of firewood sellers to be a problem. Two-thirds of them had this view, against 40 per cent of sellers in the other cities.

Transporters

As independent participants in the supply system, transporters appear to be more active and larger in scale in Dar es Salaam than in either of the other cities, though they are not a large part of the system in any of the cities.

Producers

Similarly, producers for the Dar es Salaam market appear to be larger-scale in Dar es Salaam than the other cities. This reflects the larger size of the Dar es Salaam market, and in the case of firewood shows the prevalence among users and traders in Mbeya and Shinyanga to collect or cut their own firewood to sell or use. It is also consistent with results showing informal sector traders overall to be larger-scale in Dar es Salaam (Hosier 1992).

System Integration

Participants who own and control the woodfuel from the time it is harvested to when it is sold to the end user characterize a vertically integrated woodfuel supply system. They also own the capital equipment, in particular the means of transportation, necessary to produce and sell the woodfuel. The survey conducted here shows the woodfuel supply system in Tanzania as a whole to be only moderately integrated. Full integration includes traders, who are involved in the path of the woodfuel from production to sale to the end user, as a few traders in Mbeya and Dar es Salaam in fact are. Such a system is the extremely well integrated woodfuel supply network that has existed in Sudan (Leach and Mearns 1988). This system realizes efficiencies such as coordination of transport, so that trucks carrying fuelwood into the cities carry goods out of the city on the first leg of their journey, instead of going out empty as in the Tanzanian case. It also features storage facilities for inventory of in-season purchases for off-season sales and credit facilities to carry some labourers and traders through the off-season. Despite the relatively few traders controlling the market, the study 'found little evidence of collusion' (Leach and Mearns 1988, p. 273).

Benefits of Vertical Integration

Vertical integration of the market can eliminate margins at each stage of the system, and allow unprofitable sections of the network to be offset by more profitable ones for integrated traders, thereby improving the overall system efficiency. Vertical integration in firms is generally a response to imperfections in the market, which make it difficult and costly to subcontract with others in the supply system (Williamson 1975). In the case of the woodfuel supply system, vertical integration may be a response to unpredictable supply, particularly in the rainy season when farmers are busy in their fields. It might also be a reaction to transport constraints.

Costs of Vertical Integration

On the negative side, total integration can lead to oligopoly or monopoly, depending on the size of the market being served, and implies a large capital investment. This can serve as a barrier to entry, especially since capital markets are weak and financing similarly difficult to access. Unless there is an assurance against collusion in the oligopoly scenario, reduced competition can offset any gains in consumer welfare effected by price reductions due to efficiency enhancements. Fully integrated systems can also curtail small-scale entrepreneurial endeavour, in particular the farmers who produce charcoal and harvest firewood in the non-agricultural season, thus eliminating an important source of rural income in some areas. If the integrated suppliers use full-time charcoal producers, most of whom were not members of the local community, the money generated by the woodfuel trade would not benefit the rural population. Taxes imposed and collected by local authorities, and local control over cutting rights and other land use questions would mitigate this situation somewhat. As well, local farmers could potentially band together and market their own product.

Ratio of Value Added to Total Sales

Judging solely by the number of nodes in the market chain, Shinyanga seems to be the most integrated of the three urban areas and Mbeya the least. However, because the differences in the number of nodes in the cities are not significant, another test is needed to be able to determine which city has the most integrated system. An additional test can be made by comparing the ratio of value added to total sales in each city (Perry 1989).[2] Table 4.12 shows the ratios. The results of this test show that in all seasons, Mbeya is the city with the highest ratio and therefore the most integrated system, followed by Dar es Salaam and finally Shinyanga. Mbeya's seemingly restrictedly competitive system seems to support the idea that integration can support such constraints, as well, though it is not yet clear that this is truly the case in Mbeya. Although Shinyanga and Dar es Salaam had slightly shorter market chains, the fact that Mbeya's wholesalers control the production side of the system for charcoal make their system more integrated, despite the fact that the wholesalers tend to give up control of retail sales.

Table 4.12 Ratio of value added to total sales

	Yearly average	Dry season average	Wet season average
Dar es Salaam	0.30	0.28	0.31
Mbeya	0.39	0.40	0.39
Shinyanga	0.23	0.22	0.24

While Mbeya's ratio does not change between seasons, in Dar es Salaam and Shinyanga the ratios increase during the wet season. This implies that excess profits are being captured in those towns during the wet season because supply is more concentrated and costs in the wet season are not going up as rapidly as profits. This will be explored further in the section dealing with prices.

Buying and Selling Characteristics

Entrepreneurship

One of the perceived inconsistencies in African markets cited in the introduction was lack of entrepreneurship in market participants. In the case of Tanzanian woodfuel markets, most participants claim a profit motive. Referring back to Table 4.5, most sellers are in the business for their primary income, and one-quarter to one-half feel that it is their best available source of primary income. This indicates that these participants are interested in profit. Though profit may not be the only motivation for participation in the woodfuel, or any other market system (cf. Jolly 1989), it is an important one. Most sellers, particularly retailers, complain of small profits, but this is merely an indication of their similarity to small business people everywhere.

Differentiation between Buyers and Sellers

Another structural characteristic of the market is the extent of differentiation seen between sellers and products by buyers. Woodfuel users seem relatively indifferent to seller differentiation in many cases. Particularly in Dar es Salaam, many respondents indicated that they went to any of the available woodfuel sellers in a particular market or neighbourhood to purchase fuel. When asked why they bought fuel from a particular seller, three-quarters of respondents cited convenience to home as one of the reasons, and many of those end users found that to be the most important factor in their choice of sellers, as shown in Table 4.13. This is the most compelling reason to choose a particular seller for most end users, since fuel is fairly bulky, and because many buyers buy every day and so prefer not to travel great distances to make their purchase. About half of the end users cited the quantity of fuel received for the price as a reason they chose a particular seller, About 40 per cent of end users in Dar es Salaam, about 30 per cent in Mbeya and about 20 per cent in Shinyanga chose their seller on the basis of price. Choices made on the basis of price might be low by end users because prices do not vary much within a city or market area. In Dar es Salaam over 60 per cent and in Shinyanga half of end users reported that they saw a difference between sellers. Only about a third of buyers in Mbeya reported the same. Of those who did see a difference, over half in Dar es Salaam and Shinyanga versus less than a third in Mbeya saw a difference in price. So lack of differentiation by price seems to be a function partly of lack of price difference.

Table 4.13 End user reasons for choosing a particular woodfuel seller

	Percent agreeing							
	Dar es Salaam		Mbeya		Shinyanga		All	
	CC[a]	FW[b]	CC	FW	CC	FW	CC	FW
Reason								
Convenient to home	73.6	82.2	64.2	62.9	75.0	76.0	71.3	73.5
Convenient to work	42.4	34.2	29.6	30.0	28.9	20.0	35.1	30.5
Cheapest	37.6	41.7	28.4	38.6	18.4	28.0	29.8	38.5
Greatest quantity for price	50.4	45.6	44.4	42.8	47.4	20.0	47.8	40.8
Highest quality	65.6	59.5	38.3	47.2	42.1	44.0	51.4	52.3
Seller is a relative	4.0	2.6	3.7	1.4	2.6	0.0	3.5	1.7
Regular seller has no fuel	25.6	30.4	19.8	15.7	23.7	32.0	23.4	24.7

Notes: a = charcoal; b = firewood.

Number of Buyers and Sellers

Another issue concerns the number of buyers available to buy a product, and similarly the number of sellers selling a product. This is important, because the more buyers there are competing to buy the product, the more information the seller has about prices and the easier it is to switch between buyers. Table 4.14 addresses this question. Since more than half of Mbeya producers of charcoal always sell to the same buyer, it seems that producers there have few choices in buyers. However, many of those sellers are employees of wholesalers. This fact does more to emphasize the large size of the wholesalers and the evident control they have on the market in Mbeya, than it does to exemplify the choice of the independent producers.

On the other hand, those producers who do not always sell to the same buyer sometimes have difficulty in finding a buyer. For charcoal, almost half of Dar es Salaam producers find it difficult to find a buyer at times. In

Table 4.14 Percent of participants who always sell to the same buyer or buy from the same seller

	Percent of participants			
	Dar es Salaam	Mbeya	Shinyanga	All
Producers				
Charcoal	10.6	52.0	7.7	18.8
Firewood	6.3	0.0	0.0	3.4
Sellers				
Charcoal	8.8	19.7	26.8	15.4
Firewood	16.3	2.7	22.2	11.2
End users				
Charcoal	33.1	37.8	59.2	41.4
Firewood	53.4	57.5	81.8	59.8

Shinyanga, fewer than 30 per cent feel that way, but in Mbeya two-thirds find it to be the case. So for independent producers in Mbeya there is little choice of buyers, because of the difficulty of finding one, and the evident control by the larger wholesalers. For firewood, the numbers are similar but slightly lower in all but Shinyanga, where over 70 per cent find difficulty selling their product. This would indicate that such producers are at a bargaining disadvantage, though the difficulties may be only seasonal. In other studies, there has been a concern that a poor transportation system could cause such a situation, but, in Ghana at least, few farmers felt that the situation held in the case of produce sales (Nyanteng and van Apeldoorn 1971, Southworth et al. 1979). In Dar es Salaam, these results are probably due to an oversupply of charcoal in the dry season, rather than to a lack of wholesalers. In Mbeya, most charcoal producers are tied to a particular buyer, and those who are independent have to work harder to sell their fuel. This supports the idea of a constrained wholesale market in that city, but it could just as easily be attributable to a transport constraint. In Shinyanga, there are occasional shortages of charcoal but enough wholesalers, and so producers are able to sell what they produce without trouble. Another aspect of the Shinyanga market is the fact that it overlaps with the markets for Mwanza and Tabora. This means that producers can sell to wholesalers for other cities, as well, and so there is basically always a market. For firewood, however, there seems to be evidence for a lack of wholesalers in the market. Additionally, a declining demand for firewood could be restricting the purchases of firewood from producers.

Sellers of charcoal tend to buy from a number of suppliers. Shinyanga has the highest number of sellers who are committed to a single supplier, perhaps because the small size of the market means that there are fewer wholesalers and producers. However, this commitment to a single supplier only encompasses about a quarter of all sellers, so it is difficult to determine whether this is a competitive restriction.

Over 70 per cent of Dar es Salaam retailers considered fuel supply to be a problem. None of the Dar es Salaam retailers considered it one of their most important problems, unlike 20 per cent of Mbeya retailers and about half of retailers in Shinyanga. For all, the problem is more acute in the wet season. On the other hand, 80 per cent of wholesalers in Dar es Salaam had the same complaint about supply, so this seems not to imply a lack of choice, or control by wholesalers, but rather that supply is difficult in the rainy season. This is also borne out by anecdotal evidence, and by the fact that few sellers are stuck buying from one supplier.

End users of charcoal are more likely to buy from the same retailer every time. This is particularly true in Shinyanga, where almost 60 per cent of buyers purchase from only one supplier. The size of the market again dictates fewer total sellers, and this lack of choice can pose a competitive constraint on the market. Additionally, as Table 4.15 shows, end users in Shinyanga are particularly concerned with the convenience of the seller.

Ease of Entry

Another aspect of the structure of the woodfuels market is the ease of entry by players at all levels of the system. Equal access to participation in the system is important to the competitiveness of the system because unequal access would give an unfair advantage to some players, and encourage noncompetitive behaviour. Ease of entry includes freedom from excessive capital requirements, access to credit, freedom from undesirable discrimination and collaboration preventing new entrants from doing business (Sosnick 1968).

Access to Credit

One of the determinants of ease of entry is access to credit by participants. Credit is important to the functioning of the system because it allows potential participants to enter into the system, allows current participants to expand their involvement, and allows participants to dampen seasonal and annual market fluctuations. A lack of credit availability is one of the commonly cited constraints on competitive behaviour in developing country market systems (cf. Saul 1987, Norvell and Thompson 1968, Jones 1972, Jolly 1989).

Table 4.15 shows that access to credit is somewhat uneven. It is most available in Dar es Salaam, largely in the form of prepayment of producers by sellers and fuel sold on credit to end users. Sellers have the least opportunity for credit, though they are the group that has the highest capital costs, in general. Shinyanga is the city with the least developed credit system for woodfuel traders, with no one in the sample having either taken a business loan or received prepayment for their product. Producer credit in the form of a supply of food and tools is prevalent in Mbeya for charcoal producers employed by wholesalers. This type of in-kind payments is not included in this table. It is the half of producers who are independent and do not have access to payment in kind that have little access to credit. This lack of credit access probably inhibits entry into the woodfuel business, in particular at the wholesale level.

Table 4.15 Access to credit for participants who sell to more than one buyer

	% of participants getting credit		
	Dar es Salaam	Mbeya	Shinyanga
Producers			
Charcoal	14.5	5.9	10.5
Firewood	12.5	0.0	0.0
Sellers			
Charcoal	4.5	6.1	0.0
Firewood	0.0	0.0	0.0
End users			
Charcoal	37.3	27.8	26.7
Firewood	40.6	12.5	11.1

Capital Requirements

Despite the lack of access to credit, the competitive precept against excessive capital requirements appears to be met in the woodfuel business. Though the capital requirements required for setting up business in the fuelwood markets is not known precisely, some calculations can be made with the help of assumptions. At the retail level, all that is really needed is a bag of charcoal or some cut or gathered firewood. Charcoal costs about Tsh 600–700 (US$2–2.25) per bag,[3] and firewood varies between the labour costs of collecting it and Tsh 12 (US$0.04) per kilogram. Licenses are often not purchased, but even when they are the price is minimal. About half of the retailers in the three cities also pay for the use of a space to sell their wares. More firewood sellers in the three cities, particularly in Shinyanga, pay this type of fee. Only about a third of charcoal sellers in Mbeya pay for their selling space. The average monthly fee paid by sellers in the three cities is Tsh 1200 (US$4), though the average in Shinyanga is somewhat higher at about Tsh 1500 (US$5). Firewood sellers in Mbeya, who are mostly very small-scale, female sellers with non-permanent market selling spots, pay an average of about Tsh 500 (US$1.70) per month. These fees translate to less than Tsh 0.01 per kilogram of fuel, so they can be considered virtually negligible.

At the producer level, a shovel, a hoe and a machete are sufficient capital equipment. A license is also needed for legal production, and those who cannot get one produce illegally or produce under the license of a wholesaler. Licenses are not particularly expensive, but are often constrained in number, and therefore difficult to procure. There does not seem to be an active black market in licenses, but payments and connections in the right places are considered to have positive effects on speeding their acquisition.

Of the participants in the woodfuel market chain, wholesalers are required to have the most capital to get into the business. They must be able to buy a truckful of fuel, pay for the transport of the fuel into the city, and pay potential fees and taxes on that fuel. The total amount of money required can easily be Tsh 70,000 (US$230) or more, a large sum in a country where an average monthly government salary is in the Tsh 10,000 range. Although these capital requirements keep out an 'average' person, the amount of money required is not extraordinarily high for a business person. Most wholesalers owned another business before moving into the fuelwood business, and acquired their capital in that way.

Licenses, as mentioned earlier, are the highest barrier to entry in Mbeya and also in Shinyanga. Though illegal producers abound in both areas, the rationed supply of licenses restricts both wholesalers and producers from participating fully in the system, or adds the costs of being illegal to the total costs of producing or selling.

Longevity

Finally, one way of judging the ease of entry into the system is to look at the distribution of participant longevity. From Table 4.16 we see that the greater part of participants in the woodfuel supply system have been in the business for less than five years. Producers have been in the business longer than sellers, which is logical given that they are more likely to be part time participants. This would imply a high skill level for producers of charcoal, and perhaps a corresponding increase in the production efficiency of these more experienced producers. This issue will be explored in a later chapter. On average, wholesalers have been in the business longer than retailers, except in Shinyanga. All of the wholesalers surveyed in Shinyanga have been in business for less than five years, evidencing either rapid turnover in the business or the rapid expansion. In Mbeya, more wholesalers have been in business longer than in the other towns, indicating a slow turnover or, perhaps, barriers to entry in that town. Alternately, a contraction of the woodfuel business could be happening there, but other evidence indicates that this is not the case.

Table 4.16 Seller longevity

	Mean (years)	% in business 2 years or less	% in business 5 years or less
Dar es Salaam			
Wholesale	3.2	40.0	74.3
Retail	2.7	50.0	86.0
Mbeya			
Wholesale	4.0	27.3	63.6
Retail	2.2	60.9	92.4
Shinyanga			
Wholesale	2.3	55.8	100
Retail	2.7	42.9	81.4

Summary

This chapter has described the structure of the woodfuel markets in Tanzania. They have been seen to be variable, both between fuels and between cities. A vigorous charcoal market and a slightly less competitive firewood market characterize Dar es Salaam. Experienced and independent producers and numerous independent wholesalers and retailers characterize the supply region serving Dar es Salaam. There are no indications of any particular problems in the structure of the woodfuel supply system in Dar es Salaam.

Large sellers in its charcoal market, particularly at the wholesale level, and perhaps overly numerous sellers in its firewood market characterize Mbeya. The existence of woodlots in and around the town has an affect on consumption patterns, and firewood is a proportionately more important fuel here than in the other urban areas studies. Firewood sellers generally collect their fuel to sell, and are very small scale. Charcoal producers are likely to be professional producers, and are not likely to be from the local area. They usually work under a wholesaler's license are required to sell their product to this wholesaler. Because of this, Mbeya woodfuel distribution system appears to be the most vertically integrated of those in the three cities. The root cause of this, and a severe barrier to entry into the wholesaler or production level of the woodfuel business in Mbeya, is the rationed supply of harvesting licenses. These indications in the structure of the woodfuel distribution system in Mbeya point to a system that may be less than completely competitive.

Shinyanga has a fairly evenly balanced charcoal market structure, but seems to be dominated by large sellers in its firewood market, and to a lesser extent at the wholesale level in its charcoal market. Supply problems caused by local deforestation affect the consumption and distribution characteristics of the town. Fuel travels longer distances to the town than fuel going to Dar es Salaam or Mbeya. Producers are often professionals, but less often than in Mbeya, and they are less frequently required to sell their product to a particular trader. However, retailers seem somewhat restricted in their choice of supplier, and seem to be more likely to be tied to a wholesaler than those in the other cities. Additionally, there are some barriers to entry posed by restricted harvesting licenses, but these restrictions are less severe than in Mbeya. Nonetheless, there are enough anomalies in the structure of the woodfuel supply system in Shinyanga to raise suspicions as to its competitive nature.

All three cities are plagued by little credit access and are perhaps hampered by it at the wholesale level. The lack of access to credit is particularly acute in Shinyanga.

Potential Impacts

This chapter has identified several areas of potential problems. It is difficult at this venture to predict what their impact is on the system as a whole. The next chapter moves on to the next set of criteria, examining the conduct of the sellers in the woodfuel markets in the three cities. The information learned and the problems identified in the structure of the woodfuel markets in each city will be built upon. Potential problems may be shown to have little impact, as the conduct of the sellers does not reflect the consummation of the problem. Or, problems may turn out to have serious impact, as reflected in the conduct of the participants. Finally, the problems may not show up in the conduct of the participants, but become apparent when the performance of the market is analyzed.

Notes

1. A comparison with the 1986 World Bank survey for Dar es Salaam shows very similar values. The 1986 survey found households using 1,200 kilograms of charcoal and 1,400 kilograms of firewood annually. About 88 per cent of the charcoal was used for cooking, with the residual being primarily used for ironing; 46 per cent of the firewood was used for cooking, while the rest was used for microenterprises such as beer brewing (World Bank 1987).
2. Adelman (1955) proposed this measure, 'which can be useful in comparing the extent of integration between similar firms in a given industry at a particular point in time' (Perry 1989, p. 238).
3. The 1990 parallel rate if Tsh 300/US$ is used instead of the official rate of Tsh 196/US$.

5 Conduct of Market Participants

Overview

The analysis of market conduct is the next level of evaluation in the investigation of the competitiveness and efficiency of the fuelwood market system in Tanzania. Conduct of participants in the market system refers to their behavioural aspects. In the analysis, we will look at the way in which prices are formed, the informal connections between participants, and sales promotion by sellers. The purpose is to determine if these aspects are performed in an effectively competitive manner. Constraints to competition are assumed to exist if connections between participants are not justified by cost, condition, or competition, or if price setting is done with intent to pre-empt or discriminate against competitors, or are made as part of agreements which needlessly undermine efficiency (Sosnick 1968).

Price Determination

The way that prices are set by sellers, and the degree of market and price communication among sellers are parameters in the determination of competitiveness. These aspects of price determination will help to ascertain the existence of collusion between sellers or buyers, and its role in price coordination. Additionally, the role of sales promotion in the determination of price is explored.

Price Setting

Table 5.1 presents the primary factors in setting the selling price of the fuels. Most sellers report that they set prices based on their total costs or on their purchase price of the fuel. However, in Mbeya many sellers admit to setting their prices based on the prices of other sellers of the same commodity, or to actually colluding with other sellers to set prices. Price matching is especially common among firewood sellers and charcoal sellers in Mbeya, and price collusion is particularly prevalent among Mbeya firewood sellers. This is not unexpected in the firewood sellers, since they are already extremely marginal and numerous, and cannot afford to compete intensively with each other. The level of price matching among charcoal sellers does not reflect the fact that Mbeya wholesalers who are members of the 'cooperative' freely admitted elsewhere that they do sell at the same price as other members. However, that price may be determined on the basis of total costs, and because the cooperative answered as a single being despite being responsible for 20 per cent of all sales. As a whole, Mbeya charcoal sellers look less constraining on the competitiveness of the system than was indicated in the previous chapter, though firewood sellers look more so.

Table 5.1 Primary factor determining selling price of woodfuel sellers

	Wholesalers			Retailers		
	DSM	MBY	SHY	DSM	MBY	SHY
Charcoal (%)						
Total costs (+profit)	40.6	70.0	33.3	56.1	25.0	57.1
Purchase price of fuel (+profit)	43.8	10.0	0.0	35.4	28.6	17.1
Match price of other sellers	9.4	20.0	16.7	4.9	30.4	11.4
Collude with other sellers to set price	6.3	0.0	50.0	3.7	14.3	14.3
Negotiation with buyer	–	–	–	0.0	1.8	0.0
Comparison with alternative fuels	–	–	–	–	–	–
Firewood (%)						
Total costs (+profit)	60.5	16.7	77.8			
Purchase price of fuel (+profit)	30.2	8.3	0.0			
Match price of other sellers	2.3	38.9	11.1			
Collude with other sellers to set price	2.3	25.0	11.1			
Negotiation with buyer	2.3	8.3	0.0			
Comparison with alternative fuels	2.3	2.8	0.0			

In Shinyanga, some sellers also indicate that they base their prices partly on the prices of other sellers, or in collusion with other sellers. Collusion is reported most often among charcoal sellers, who admit to setting prices with other sellers in half of the cases. Only a third of the wholesalers report setting their prices based on total costs. These indications imply that the wholesalers in Shinyanga may be impeding the competitiveness of the system in that way.

Charcoal retailers in Mbeya are more likely than those in the other cities to set their prices based on either other sellers' prices or in collusion with other sellers. They are also much less likely to use total costs as a basis for their price setting. Sellers of all kinds are most likely to use total costs to set prices in Dar es Salaam. Costs plus profit, or mark-up, is the method most likely to be identified with a 'competitive' system, and the use of the purchase price of the fuel plus a profit percentage is the next. Sellers in Dar es Salaam report these methods of setting prices most frequently, where at least 80 per cent of all sellers claim to use them. Except for charcoal wholesalers, these methods are least used in Mbeya, where as few as 25 per cent of sellers use them for setting firewood prices, as compared to over 90 per cent in Dar es Salaam.

Price Communication

In general, there appears to be a relatively high degree of price communication, among sellers in the woodfuel markets. What is not clear is whether any or all of the sellers calculate their prices in reverse, by 'knowing' the going retail or wholesale price and calculating their supply purchase price based on this.

This question can be partly addressed *via* the information contained in Table 5.1. Shinyanga sellers are less likely to set their prices below those of other sellers, and more likely to avoid doing so in order to escape alienating other sellers, with whom they had originally collaborated to set the price. Mbeya charcoal wholesalers report themselves as being the most likely to set their prices lower than their competitors, and say that is due to lower costs for themselves, or that it is purely for the sake of sales promotion.

Charcoal retailers in Shinyanga are the most likely, and those in Mbeya the least likely, to set their prices lower than other sellers. As was the case for wholesalers, retailers report that the lower prices are set because of lower costs or for sales promotion.

Firewood sellers in Mbeya are also the least likely to lower their prices below those of other sellers. The larger firewood sellers in Dar es Salaam and Shinyanga are more likely than the small sellers in Mbeya to lower their prices. Again, avoiding conflicts with other sellers is the most common reason quoted for avoiding this practice in all three cities. For those who do lower their prices below their competitors, the largest number do so in order to attract customers. Unlike charcoal sellers, few firewood sellers lower their prices due to reduced costs for themselves. This coincides with the information found in the previous table. Because firewood sellers are marginal, earn little profit, and set prices to match those of other sellers, they are therefore very unlikely to undercut prices.

Fuel Quality

Another reason for setting prices below those of a competitor's, according to Table 5.2, is when a seller has low quality fuel. As seen in the previous chapter, in Dar es Salaam almost two-thirds of end users choose a seller based partly on the quality of the fuel they are selling, and some buyers found this to be the most important consideration when selecting a seller. About 40 per cent of buyers in the other two cities were picky about fuel quality. Cline-Cole (1984, 1987) has written extensively about product differentiation in woodfuel markets in Sierra Leone. There, he has found market preference to be widespread and complex influential on sellers and producers. It is 'common knowledge' there and in Tanzania that charcoal which is 'light' when picked up is of lower quality (in terms of burning time) than charcoal which is 'heavy'.[1] Sellers are even more cognizant of fuel quality. Over three-quarters of sellers will reject low quality fuel if offered to them by their supplier. If nevertheless stuck with low quality fuel, most retailers will offer it at a discount to their customers, or give a larger quantity of the fuel for the same price. More retailers offer this concession to their buyers in Dar es Salaam and Shinyanga than in Mbeya (77 per cent and 69 per cent versus 57 per cent). Some retailers who do not might be constrained by suppliers. Small retailers seem to have some trouble with fuel supply, and when a supplier delivers sub-par fuel she feels she can not refuse it without losing her source of supply. However, this is not the general case.

Wholesalers are less generous with their low quality fuel. In Dar es Salaam, fewer than a quarter of wholesalers give some sort of price or quantity discount for low quality fuel, while half in Mbeya and two-thirds in Shinyanga give such discounts. This is non-intuitive, since more retailers sell the low quality fuel at a discount in Dar es Salaam, and this means that sellers then have to reduce their margins in these cases. This might best be explained by the fact that buyers there are the most particular about quality of the three cities, and so retailers are forced to compensate for that. Retailers also have more trouble attaining fuel supply in Dar es Salaam than elsewhere.

Table 5.2 Sales promotion by sellers

	Wholesalers			Retailers		
	DSM	MBY	SHY	DSM	MBY	SHY
Charcoal (%)						
Set prices below other sellers'	15.6	44.4	0.0	23.5	10.7	48.6
Why?		All cities			All cities	
To attract customers		37.5			32.5	
When get fuel or transport cheaper than usual		50.0			32.5	
When have low quality fuel		0.0			17.5	
To have quick sales for money/new stock		0.0			12.5	
For special customers		12.5	5.0			
Why not?	DSM	MBY	SHY	DSM	MBY	SHY
Avoid conflicts with other sellers	35.7	0.0	60.0	58.5	76.7	66.7
Small profits/lead to losses	64.3	100	40.0	41.5	23.3	33.3
Firewood (%)						
Set prices below other sellers'	28.6	17.1	33.3			
Why?		All cities				
To attract customers		35.3				
When get fuel or transport cheaper than usual		17.6				
When have low quality fuel		17.6				
To have quick sales for money/new stock		17.7				
For special customers		11.8				
Why not?		All cities				
Avoid conflicts with other sellers		60.9				
Small profits/lead to losses		39.1				

Relationships between Buyers and Sellers

Only a few sellers in Table 5.2 indicate that they lower their prices for certain buyers. In Table 4.13 it was seen that few end users choose their supplier on the basis of family ties. Similarly, only between 1 per cent and 2 per cent of sellers choose their fuel suppliers because they are either family members or from the same village. A similar number of producers sell their product exclusively to their relatives. Additionally, fewer than 10 per cent of producers outside of Mbeya sell to an exclusive buyer, though a higher percentage of sellers in Shinyanga and Mbeya sell to an exclusive buyer. Exclusive exchange relationships, even if voluntarily entered, may segment the market in the face of several traders, thereby creating an element of monopolistic competition. They serve to inhibit the increase of prices in response to supply conditions (Saul 1987). This is a problem in Mbeya and among the wholesaler-retailer connection in Shinyanga.

Credit

There does seem to be a perception by end users that sellers from whom they get credit are different from those by whom they are not given credit. In Dar es Salaam and Shinyanga, end users see those who give them fuel on credit as being less expensive. In Mbeya households see the quality as being better from credit-giving suppliers, and in Shinyanga end users who buy on credit feel that the quantity they are given is greater. If actual prices and quantities are examined, these perceptions are shown to be untrue.[2] The quantity given to retail buyers of charcoal is the same for those who do or do not buy on credit. The price differences between fuel bought on credit and that bought with cash are generally insignificant, with two exceptions. The first is the price of bag (bulk) sales of charcoal in the wet season.[3] During the wet season, fuel bought in bulk on credit is significantly cheaper than fuel bought with cash. In this case, there is a special relationship between the participants, and those with this relationship are afforded a better price. It is not clear why this is true, since woodfuel tends to be scarcer in the wet season. Perhaps special customers get a price in the wet season that is close to the dry season price, while other customers pay the higher, wet season price.

In the case of producers selling wet season firewood, those producers who are paid in advance for their fuel get a lower price for it when they sell it. Though fuel is somewhat scarcer in the wet season, the extra value of fuel produced then is lost to producers who are paid in advance. The advance payment is a form of credit, and the lower price an interest rate imposed on the loan. This problem of special relationships and their constraining effect on competition is often bemoaned in the literature, but seems not to be a problem in Tanzania for charcoal producers, and is not particularly acute for firewood cutters.

Sales promotion The extension of credit is important to the conduct of the market in additional ways. Only information on the extension of credit by sellers is available, and it is presented in Table 5.3. Retailers are seen to be more likely than wholesalers are to give credit to buyers. Only in Shinyanga are wholesalers more likely to extend credit than retailers. Mbeya is the place where a buyer is least likely to be extended credit by a seller.

Sellers can view the extension of credit as another type of sales promotion. Rather than lowering prices, a service is provided by the extension of credit. Although there is no difference in pricing between sellers who extend credit and those who do not, when the added service and a positive interest rate is considered, financial benefit of buying from a seller on credit can make the difference in seller choice. This can be particularly significant with wholesalers who extend credit. For retailers who are just starting out, or who are marginal, finding a seller who will extend credit may be the difference between being in or out of business. Credit relationships facilitate trade, and provide traders with commercial opportunities they could otherwise not take advantage of.

Table 5.3   Extension of credit by sellers

	Charcoal (%)		Firewood (%)
	Wholesaler	Retailer	
Dar es Salaam	34.4	48.8	37.2
Mbeya	30.0	37.5	13.5
Shinyanga	66.7	48.6	22.2
Total	37.5	45.1	25.8

Shinyanga

The large number of wholesalers extending credit in Shinyanga may be connected to the fact that the charcoal traders in that city are larger than those in the other cities. Additionally, no Shinyanga wholesalers reported price-cutting as a form of sales promotion, so the extension of credit may be seen as a substitute for this. Conversely the large number of credit sales may preclude the possibility of price competition. When combined with the fact that over a quarter of Shinyanga sellers sell to an exclusive buyer, there are indications that some sort of 'special relationship' is relatively common between wholesalers and retailers there. This may be connected with a 'sticky' price response, and this will be explored in succeeding chapters.

Firewood

For firewood, the overall percentage of traders extending credit is lower, but the highest percentage exists in Dar es Salaam, followed by Shinyanga. In Mbeya, little credit is given by firewood sellers, which is to be expected in two ways. Firstly, sellers are mostly, as discussed, extremely subsistence oriented, and are unlikely to have the wherewithal to extend credit. Any moneys collected are likely to be needed to provide daily provisions for the seller's family. Secondly, a greater percentage of the Mbeya population uses firewood, so buyers are numerous and relatively small-scale. This contrasts with buyers in Dar es Salaam and Shinyanga, who are more likely to be involved in a home industry such as beer brewing, and are in any case in shorter supply than in Mbeya. The sellers in Dar es Salaam and Shinyanga are also much larger on average than those in Mbeya, and so have more resources to draw upon.

Summary

It seems that special relationship between buyers and sellers are not significant in the Tanzanian fuelwood market, at least as is reported by the buyers and sellers themselves. Although in some places, in particular Shinyanga, credit extension by sellers is common, it is not clear that this has engendered any particularly competition-constraining behaviour.

It also has been seen that in some cases, especially in Shinyanga among charcoal wholesalers and Mbeya among firewood sellers, that prices are set in noncompetitive ways, ways that we presume may result in higher prices for consumers, and therefore a welfare loss for that group. The results of this conduct by sellers will be examined in following chapters, to determine if these presumptions are correct and if, in fact, consumers are hurt by this fault in market conduct.

Overall, the research seems to support the idea of the role of personal bonds in African markets. Although personal bonds are not everywhere evident, in some cases at least they have an impact on the functioning of the market.

Notes

1. This is a rule to take into account the physical properties of wood: a piece of wood with a high density (more kg/m^3) will feel heavier than one of the same size with a low density. If the wood is dry, a high density, or heavy, piece of wood will have a higher heat content than a low density piece. Wood with a higher heat content will burn longer or hotter than wood with a lower heat content, and is therefore preferable for most household uses. The rule breaks down when the pieces of wood are not equally dry.
2. The results of the t-tests comparing selling or buying price of fuel for credit givers vs. non-credit givers are as follows:
   *   a kopo is a commonly used charcoal measure of about 1 kg;
   ** significant at the 90 per cent range or better.
3. This is actually the wholesale price of charcoal, the price faced by retailers buying fuel to sell. Most retailers buy fuel from wholesalers, and the fuel is bought in bulk, in bags.

6 Market Performance – Productive Efficiency

Overview

The previous chapters have examined the structural and conduct aspects of woodfuel markets in Tanzania. The remaining aspect to be examined under the SCP paradigm is the performance characteristics of the market system. Market performance is the economic result of structure and conducts as measured in terms of prices, costs, distributive margins, and output. Performance is the reigning element in the SCP structure, and it is from the exploration of performance that the most illuminating information is retrieved.

In this exploration the question of efficiency is paramount. The objective of a market system is to facilitate price formation. In turn, the pricing system has as its objective guiding the flow of resources into production and market, and of goods and services into consumption (Bressler and King 1970). An efficient market system will provide 'efficient and economical services and ownership transfers in the movement of commodities from seller to buyer, and an effective price-making mechanism' (Bressler and King 1970). These two aspects, the productive and pricing efficiencies of the market system, will be explored separately, with performance efficiency being considered in the present chapter and pricing efficiency in the following chapter.

In looking at the productive efficiency of the woodfuel market system, the extent to which participants take advantage of economies of scale; the input/output ratio of participants of different sizes and locations; and the extent to which participants in the market fully utilize their capacity will be examined. Capacity, or load factor, in a business such as the production and market of woodfuel is not a straightforward subject. Determining whether market participants make 'reasonably full use of their available facilities' (Bressler and King 1970) is difficult in a market which is informal and, in some cases, marginalized. However, an attempt can be made to discuss, with available data providing support, where and to what extent capacity is an issue in the woodfuel market.

Technical Efficiency

For charcoal producers, the scale factor will be actualized in three ways, by kiln size, by labour invested in a bag of charcoal, and by total sales. The first is the more theoretically based, and is dependent on physical efficiencies as well as experience and knowledge.

Kiln Efficiency

In Tanzania, traditional earth-covered kilns are used almost exclusively in the production of charcoal. Although a recent project has attempted to introduce improved kilns such as the Casamance kiln to some areas in Tanzania, notably the supply areas for Dar es Salaam, few charcoal makers have adopted improved kilns for regular use.

Traditional kilns vary in both size and shape in different regions of the country. In general, kilns are rectangular, but in some areas, particularly in the Shinyanga supply area, circular kilns are used. Most of the work on theoretical efficiency relates to circular kilns, but the general conclusions can be applied to rectangular kilns as well. In general, the larger the kiln, the greater the physical efficiency of the kiln (Feinstein and van der Plas 1991). However, an economically efficient kiln is limited by efficient labour utilization, and is lower than the physical optimum size (Karch 1987). Field tests in Rwanda, Jamaica and Senegal have confirmed the increase in kiln efficiency with size, but the size of kilns constructed by charcoal makers are dependent on a number of factors. These include access to wood and preparation tools, organization of labour, ability to have money tied up in a kiln for a longer time (a large kiln takes longer to stack and burn), kiln management skill, and risk aversion (the flare up of a large kiln represents a greater loss) (Feinstein and van der Plas 1991). In general, it is assumed that these factors are utilized to the best of the producer's ability, and the constraint on capacity is the limit of the least available of these factors.

Scale and Allocative Efficiency

Producer Efficiency

Table 6.1 tabulates a number of factors relating to charcoal producers in the country. Included there is the size distribution of kilns in areas supplying the three cities in this study. Kiln size in Mbeya is appreciably larger than that in Dar es Salaam or Shinyanga. Because wholesalers organize much of Mbeya's production, larger work teams are assembled to cut, stack, and burn kilns, thereby allowing larger kilns to be made. Professional charcoal makers also have greater access to tools provided by their sponsors, and no worries about risk to invested capital. These factors are all of greater importance to independent producers, such as those in the other two cities.

Table 6.1 Efficiency factors for charcoal producers

	Dar es Salaam	Mbeya	Shinyanga	All
Bags of charcoal per kiln*	58.4	108.8	68.5	71.3
Bags of charcoal per year	1397	1031	887	1218
Man-hours of labour per bag	4.1	5.5	2.2	4.0
Years in charcoal business*	10.8	3.5	2.5	7.4
Km from city*	101.6	115.4	171.6	120.0

* These numbers are significantly different between cities at the 90% level.

Operator Efficiency

For a rural producer, the overwhelmingly important input into a charcoal kiln is labour. Accordingly, the measurement that best describes operator efficiency in this study is man-hours of labour per bag of charcoal. Between cities, this measurement does not vary appreciably. However, within cities the efficiency measure is negatively correlated with kiln size.[1] Therefore, in Tanzania the economic efficiency of kilns seems to increase with size, as has been the case in other countries. This scale economy, combined with the evidence that physical efficiency is also scale dependent, argues well for policies to encourage larger kilns. It also implies that kilns in Mbeya are likely to be more efficient, both economically and physically.

Relationship between price and size The price received by the producer was not significantly related to kiln size. This relationship might be expected, due to transport and time efficiency for the wholesaler, since if he can go to only one or two producers and fill his truck, he saves time and money. However, such was not found to be the case. Price was related to the overall size of the producer, as measured in bags produced per year, rising with producer size.[2]

This may imply a closer relationship between large, consistent producers and their wholesalers than between wholesalers and less consistent producers. This is an informational efficiency, since it is easier for a wholesaler to visit a producer whom he knows to have fuel than to find a producer who rarely produces. In the same way, it is more difficult for an infrequent producer to find a wholesaler to buy his fuel, and he may be less informed about the price for which the fuel is selling.

Total output On a total output basis, scale efficiencies are not realized. Total yearly charcoal output was not significantly related to the defined efficiency factor, hours of labour per bag. This finding is unexpected because someone who set up to produce charcoal regularly might be expected to be more efficient than someone who only infrequently produces. Experience is also found to be unrelated to the efficiency measure. This supports the finding that scale efficiencies are not realized on a total output basis.

A negative relationship was found between experience and the distance a kiln is located from a city.[3] This tells us only that those producers closer to the city are more likely to be regular producers, since they are more able to sell their product. They also get a higher price for their charcoal than do those further away from the city.

Firewood harvesters realize scale factors only *via* total sales. This data shows no significant relationships between total output and total input, measured again as hours of labour spent cutting and market fuel. The measurements of both input and output are somewhat questionable, so no reliable conclusions about scale factors are possible.

Seller Efficiency

Sellers, both wholesalers and retailers, are subject to both scale and load factor limitations. The availability and use of labour and storage define load factors, but scale factors are measured by the relationship between input and output and the differences between differently sized sellers.

Storage Prices will be discussed in detail in the following chapter, but for now we will say that Tanzania follows other developing countries in the seasonality of its woodfuel prices. Prices are higher in the wet season due partly to higher costs and potentially also due to higher profits. One fairly obvious way to improve profits and efficiency for sellers of woodfuels would be to arbitrage and buy extra fuel in the dry season, at low cost, and store it until the wet season when it will fetch a significantly higher price. This has been suggested to be the norm in many developing country food market systems. However, very few woodfuel sellers do this – only 12.7 per cent (unweighted) of those surveyed. This lack of storage seems on the surface to be an inefficiency, a failure of sellers to make full use of their available opportunities. Those who do store, store up to 600 bags of charcoal at a time, with an average of 82 bags. However, this is only about 5 per cent of their wet season needs, so even those who store do not store in large quantities relative to their needs. The only group that seemed to store in some quantity was the cooperative group in Mbeya. This may be possible for them because of their shared labour force and capital base.

Table 6.2 Capital turnover and days between supply purchases – charcoal traders

	All	Dry season Whole-salers	Retailers	All	Wet season Whole-salers	Retailers
Capital turnover (days)	6.1			5.1		
Dar es Salaam	7.0	4.9	7.8	5.0	2.6	5.9
Mbeya	3.3	2.0	3.6	3.6	1.6	4.0
Shinyanga	4.3	2.5	4.6	6.9	3.0	7.6
Days between supplies	11.0			16.9		
Dar es Salaam	12.6	11.8	12.8	19.5	17.1	20.2
Mbeya	4.7	6.2	4.5	6.8	9.4	6.5
Shinyanga	19.2	18.4	19.3	27.5	29.2	23.7

Those who store charcoal do so for three reasons: to overcome fuel supply difficulties in the wet season; to offset the high cost of transport in the wet season; or, to a lesser extent, to engage in arbitrage. The reasons expressed by firewood dealers were similar, with the added justification of a higher tax bill for those who do store. It would be expected then, that large sellers would be more likely to store fuel, but as it turns out those who do store are no more likely to be large sellers than those who do not. The lack of storage in the woodfuel sector seems to be mostly associated with the lack of capital as a whole. It may be argued that total system efficiency is lowered because of this, but it seems that individual producers, without greater access to credit, have little chance of expanding in this area.

Labour As is the case in many informal industries, woodfuel sellers tend to use primarily their own family labour and little paid labour. Overall, half of sellers work alone, while the other half of sellers work with at least one helper, generally a family member. Table 6.3 gives more detail. For both fuels, sellers in Shinyanga use the most labourers on average, and sellers in Mbeya the least.

Although there is no connection between number of paid labourers and scale of business, for firewood there is a significant relation between total workers and scale of business. For charcoal dealers the connection is between primary income source and labourers – those who depend on another, non-farming profession for their primary income have more labourers for their charcoal business, on average. Additionally, gross earnings are positively correlated with increased spending on inputs.

Table 6.3   Labour force for sellers

Mean number of labourers (in addition to seller)	DSM	MBY	SHY	All	Median	% with 1 or more labourers
Charcoal						
Family labour	0.5	0.3	0.1	0.4	0	35
Paid labour	0.33	0.17	0.95	0.40	0	19
Total	0.9	0.5	1.2	0.8	0.5	50
Firewood						
Family labour	0.3	0.2	0.4	0.3	0	30
Paid labour	0.98	0.27	1.7	0.75	0	28
Total	1.4	0.5	2.1	1.1	1.0	50

This pattern of labour use is due to several factors. The most important is that most sellers are operating at a scale that does not require paid labour. However, when a seller has increased his business to a larger scale he eventually comes to the limit of his labour supplies, and requires an expanded labour force in order to increase his business. Most sellers either have not reached that scale of business, or they decide to restrict their business scale because of perceived or actual lack of labour. Anecdotally, sellers are often afraid to trust a non-family member with their business, thereby placing a constraint on expansion. Instead, sellers such as the Mbeya cooperative have decided to expand vertically, and to band together horizontally to increase the number of trusted business partners. This lack of business expansion opportunities is partly attributable to undeveloped legal infrastructure in the form of a lack of contract enforcement and recourse for a businessman who is cheated by one of his employees. At times, the lack of expansion by sellers is also due to lack of capital, either start up capital or working capital with which to pay his expanded labour force.

Economies of scale Charcoal sellers encounter some economies of scale. Table 6.4 demonstrates the relative margins enjoyed by sellers of different size. Larger sellers are significantly better off on a per bag basis than smaller sellers, especially when imputed wages are included from the hours of labour per bag. However, when sellers are split into wholesalers and retailers, the wholesalers experience definite economies of scale. As well, retailers who are considered medium size do better than either small or large retailers. While the smallest retailers seem to lose money (though this may be due to some cost assumptions), medium-sized retailers do better than either similarly sized wholesalers or large retailers do. This may have to do with the fact that retailers sell much of their product in small lots, affording a greater per bag margin than the bulk sales that characterize wholesalers.

Table 6.4   Scale efficiency in charcoal sellers

Tsh (mean)	Seller scale (by number of bags sold)			
	Small (<200/mo)	Medium (200-450/mo)	Large (>450/mo)	All
Margin per bag (wet season)				
All sellers	-61.8	164.8	186.5	112.0
Wholesalers	34.1	156.2	231.6	165.9
Retailers	-80.9	168.9	141.1	86.4
Margin per bag (dry season)				
All sellers	-30.7	176.3	200.2	129.9
Wholesalers	81.7	135.4	245.7	167.1
Retailers	-56.6	194.9	151.4	111.8
Labour per bag (hours)				
All sellers	3.5	0.93	0.36	1.7
Wholesalers	3.6	0.97	0.36	1.2
Retailers	3.5	0.91	0.35	1.8

Large wholesalers and medium-sized retailers are more operationally efficient both in the labour embedded in the product and in the margins they receive. This implies that the market system would be improved by encouraging larger wholesalers. However, a corresponding issue is at what point such concentration begins to dampen competition and has a detrimental effect on prices. This measure of scale efficiency is influenced by the fact that most large wholesalers are based in Dar es Salaam, and they face the most competitive atmosphere. The larger wholesalers in Mbeya and Shinyanga fall mostly in the 'medium' category.

For small and large sellers, and medium-sized retailers, margins per bag drop during the wet season. This is especially precipitous for small wholesalers. Transport charges increase in the wet season, and small wholesalers seem to be unable to capture much of that increase in their prices. Their lack of size means that they can not increase their final price enough to cover their greatly increased costs. Large wholesalers, who seem to have more economic strength, and whose per bag costs do not increase as much as their smaller competitors because of their scale economies, are able to cover most of their higher wet season costs with price increases.

Medium-sized wholesalers' margins per bag uniquely increase in the wet season. They alone are able to take advantage of the price increases to capture excess profits. It is not clear why this is so, but it may be because they make part of their sales in small lots and some in bulk. It is also probably connected to the fact that many medium-sized wholesalers are located in Shinyanga. In the wet season, these wholesalers might have increased control over prices there, as people who would normally use self-collected firewood switch to charcoal during the rainy time.

Transport Efficiency

Transport bridges the spatial aspects of the woodfuel market system. As discussed in chapter 3, trucks are the most commonly used form of transport for moving woodfuels from the harvest or kiln site to the cities. There are many common problems with this transport system, but the largest in terms of operational efficiency is the lack of full capacity utilization. Over 70 per cent of respondents always go out to collect woodfuel with an empty vehicle, and another 15 per cent sometimes go out empty. About a quarter of them sometimes carry other cargo along with the woodfuel, typically a few sacks of agricultural produce from the area from which they collect the fuel. This complements the extremely poor condition of the vehicles, their resultant propensity towards breakdowns, and to their position in the informal sector of the economy. Since many of the products transported out to the rural market centres are imported or locally made manufactured goods, most of the goods are transported by formal sector vehicles, usually parts of fleets owned by businessmen. Though there is some woodfuel transported into the city by these trucks on their return trip, especially from roadside traders, the bulk is carried by vehicles dedicated to woodfuel transport.

Reorganization to allow back hauling This implies a potential efficiency improvement in the transport of woodfuels by organizing the system to allow the transport of woodfuels on a back haul basis, with cargo being carried on both the outgoing and incoming trips. This would cut transportation costs of woodfuel to some fraction of the total depending on the value added, and be a much more efficient use of imported transport fuels as well. Woodfuel traders and transporters in the Sudan have reportedly achieved this sort of organizational sophistication. and it is thought that this is what has lowered woodfuel prices in Khartoum (Leach and Mearns 1988, p. 218). The status of the woodfuel distribution system is unclear at this time due to the destruction caused by the recent civil unrest there.

Lack of vehicles The high cost of transport is increased by this inefficiency. A general transport bottleneck exists throughout the country, and is particularly severe in Dar es Salaam. There are simply not enough vehicles to deal with the demand for transport. The condition of the vehicles used limits their applicability toward other cargo, helping to free them for use in woodfuel transport, but the demand is high nonetheless. The high cost of importing new vehicles and spare parts also adds both to the high price charged for transport of cargo such as woodfuel, and to the final price of the woodfuel.

Vehicle size Because of the poor roads, truck size is also limited severely. The trucks used are limited to 7- and 10-ton trucks that carry 150 and 200 bags of charcoal, and a similar amount of firewood, respectively. With road improvements, much larger volume vehicles such as double trailer trucks could be used. There are economy of scales present in transportation which would ensure efficiency improvements with greater size.[4] In addition to the infrastructural improvements and capital investment required, this transition would also require reorganization in the transport of the fuel. It would be necessary to have depots where woodfuel is brought from individual production sites and then loaded onto trucks to be moved into the cities. These depots could easily mimic the roadside sales points for woodfuel already in existence along major roads leading into the cities. The smaller vehicles that are now used to carry the woodfuel to the city could be used to transport woodfuel from the rural sites to these roadside stands, and the large truck move the woodfuel into the towns from that point. This idea will be discussed in greater detail in the policy section of the final chapter.

Organizational skills These kinds of efficiencies, back hauling and the use of larger vehicles and loading depots, require organizational skills and capital beyond the previously demonstrated in Tanzanian market systems. There is no reason, however, that this goal could not be met over a number of years, as it has been achieved in the past in places such as Sudan.

Train transport The final major efficiency loss in the transport sector of the woodfuel market system is the lack of availability of trains for transport. Trains are demonstrably the cheapest mode of transport per kilogram/kilometre in the country, but poor management and lack of accessibility to users limit their use. Many of the harvesters are located on or near the train lines, and the product of others could be transported there, much as fuel is now transported to the roadside for sale. Having a number of depots where the fuel could be gathered for pick up would enhance efficiency. Now most use of trains for transporting woodfuels is illegal, with the consequent costs of risks and accompanying bribes . Changes in the rules governing the use of trains are required to be able to utilize the railways.

Summary

Productive efficiency in the Tanzanian woodfuel market system has been seen to be quite variable, in some cases evidently complete and robust, and in other cases somewhat lacking. In almost all areas, economies of scale exist, and in most cases these economies are not being taken advantage of. Lack of capital and other financial instruments are the major culprits, though risk aversion is also a potential hindrance. At the producer level, kilns are shown to have both economic and physical economies of scale, as kilns get more technically efficient as they get larger. However, a larger producer, on an annual basis does not appear to be more efficient than his competitors.

This is not the case for traders. Larger wholesalers and, to a point, larger retailers seem to be more efficient than their colleagues. Possible efficiency losses are experienced due to the failure of sellers to store fuel until the wet season, and in the failure of larger wholesalers to hire labour. Again, risk aversion and lack of available capital can be identified as reasons for these shortcomings.

Infrastructural problems limit transport efficiency, but organizational hurdles and a shortage of capital equipment also pose important constraints. It is in transport that the greatest problems in capacity utilization are met. The efficiency loss experienced through the failure of transporters to back haul is particularly costly because it involves the waste of transport fuel, procured by scarce foreign exchange.

Notes

1. $r = 0.303, p \leq 0.01$.
2. Dry season price = 22.5 - 0.209 (kiln size) + 0.008 (bags/yr)
   (-1.20)   (2.75)

   Wet season price = 261/9 - 0/307 (kiln size) + 0.009 (bags/yr)
   (-1.57)   (2.96)*

   * significant at $p \leq 0.01$.
3. Experience = 177.0 - 0.734 (km from city)
   (-0.402)*

   * significant at $p \leq 0.01$.
4. For example, in Shinyanga, a 7-tonne vehicle carries charcoal at Tsh 300 per bag, while a 10-tonne truck charges only Tsh 250 per bag.

7 Performance – Pricing Efficiency

The pricing efficiency of the woodfuel market system is the final aspect of its performance that must be investigated. Problems in the structure and conduct of the system can be traced from the behaviour of the prices in the sector. Once identified, these problems can be approached with intent to eliminate them. Some of what will be discussed below will emanate from issues approached from a slightly different angle in previous chapters. However, in this chapter the price aspects of the issues will be fully explored and linked to their cause and effect to as great extent as possible.

Spatial Issues

Price Structure

The price breakdown for a typical participant along the woodfuel supply chain in each city is shown in Table 7.1. However, there are many variations on this scenario, and in particular it must be remembered that wholesalers that sell by kopo[1] will have higher earnings by weight of charcoal sold than those who sell only by bags. The table gives an idea of returns to labour for traders and producers.

Margins In general, the margins for charcoal traders in Mbeya are substantially better than for traders in other cities. This gives credence to the belief discussed earlier that there is a less-competitive atmosphere in the market there. Though the Mbeya traders report more time spent per bag (see Table 7.1), they also report lower labour costs, taxes and fees, and transport costs. The producer margin is also the lowest in Mbeya, due to higher labour costs and higher taxes and fees, as well as a lower ex-kiln price. This finding is probably distorted by the existence of the vertically integrated large wholesalers, who absorb the producer margin as well as the trader margin, and hire labour to

Table 7.1 Fuel cost structure

	Dar es Salaam Tsh/bag	% of total	Mbeya Tsh/bag	% of total	Shinyanga Tsh/bag	% of total
Charcoal						
Typical producer's cost						
Taxes and fees	6.9	1.1	14.9	2.5	5.8	0.9
Paid labor	5.3	0.8	14.9	2.5	4.8	0.7
Producer margin	138–238	28.9	110.2	18.4	139.4	21.5
Ex-kiln price	150–250		140		150	
Typical trader's cost						
Bags, tools	38.3	5.9	46.6	7.8	63.9	9.8
Paid labor	5.4	0.8	1.9	0.3	20.0	3.0
Taxes and fees	43.1	6.6	4.8	0.8	44.4	6.8
Transport	223.7	34.4	140.8	23.4	276.0	42.5
	(2.2/bag/km)		(1.2/bag/km)		(1.6/bag/km)	
	(0.05/kg/km)		(0.03/kg/km)		(0.04/kg/km)	
Storage/rent	<.01	0.0	<.01		<.01	
Trader margin	40–240	21.5	265.9	44.3	45.7–145.7	14.7
Wholesale price per bag	600–700		600		600–700	
Retail price (kopo sales)						
Per bag	810–845		900–1080		840–1120	
(Per kopo)	(30–35)		(25–30)		(60–80)	
(Per kg)	(20.8)		(21.7)		(25.2)	
Retailer margin	110–245	(21.5)	300–480	(39.4)	140–520	(33.6)

Firewood	Tsh/kg	% of total	Tsh/kg	% of total	Tsh/kg	% of total
Typical producer's cost						
Taxes and fees	0.03	0.5	0.33	2.8	0.1	1.1
Paid labor	1.5	24.6	0.00	0.0	0.03	0.2
Producer margin	1.0	16.4	5.2	47.7	4.2	47.7
Producer price	2.5		5.5		4.3	
Typical trader's cost						
Paid labor	0.15	2.5	0.00	0.0	0.03	0.2
Taxes and fees	0.03	0.5	0.33	2.8	0.09	1.0
Transport	3.4	55.7	3.3	30.3	3.1	35.2
	(0.05/kg/km)		(0.03/kg/km)		(0.03/kg/km)	
Storage/rent	<.01		<.01		<.01	
Trader margin	0.2	3.3	1.8	16.5	1.3	14.7
Retail price	6.1		10.9		8.8	

produce charcoal. These integrated traders attributed some of the costs normally borne by the trader (in particular taxes and fees) to the producer. The combined producer-wholesaler margins in Mbeya, which might be expected to be reduced because of integration, are more (at Tsh 376.1) than the average of those for either Dar es Salaam (Tsh 190–Tsh 490) or Shinyanga (Tsh 185–Tsh 285), both of which have non-integrated systems. Evidently, the Mbeya wholesalers are absorbing any results of efficiency gains effected by system integration there. Since they are competing against some wholesalers who do not have the same cost advantages of integration, they evidently feel no pressure to reduce their wholesale prices to reflect their lower costs.

More surprisingly, typical small retailers in Shinyanga have a much higher margin than the wholesalers in that city. Wholesalers are as likely as not to double as retailers, so it is often wholesalers are actually receiving the margins of wholesaler and retailer. In general, small retailers will extract a higher rent because they deal in very small quantities, requiring a higher percentage return in order to attain a living wage. Sellers in Shinyanga and Mbeya are more likely than those in Dar es Salaam to sell charcoal as their primary income source, rather than as a side business to augment income from another source. For this reason, their margins may be expected to be, and are, higher. Transport is the largest cost component for all areas, even surpassing the producer price of the fuel. However, bags and transport, both provided by the trader, are more expensive in Shinyanga than in the other cities, as is the amount of paid labour. As a result, trader margins are lower there. The retailers in Shinyanga seem to be collecting extra profits, perhaps a response to increased demand and reduced or stable supply.

Transport Costs

The per kilometre price of transport is the highest in Dar es Salaam, probably because of competition with other uses for the vehicle and general transport supply shortages. However, most traders did not complain of difficulty in finding transport. In Shinyanga, the transport problem results from it being a small town with relatively little commercial traffic, and probably fewer second- or third-hand trucks available for use as woodfuel transport. Mbeya benefits from being the hub of Tanzania-Malawi-Zambia trade, and from being an active agricultural cash crop area.

Cost Structure

Figure 7.1 and Figure 7.2 show the differences in the cost structure of the fuels graphically, highlighting the differences between the cities. It is clear that Mbeya traders are taking a relatively large percentage of the final fuel price as a margin, and it is similarly clear that Shinyanga's transport costs are reducing the percentage profits of its wholesalers. Profits in Dar es Salaam are evenly divided between transport, trader's margin, and producer's margin. In a previous chapter it was seen that in Dar es Salaam the producer margins are the largest component of final price and the producers were found to be independent. In Mbeya, the producer margins occupy the smallest proportion of final price, the producers are mostly wholesaler controlled. In Shinyanga, both the percentage of final price commanded by the producers and the structure of the production system fall somewhere in-between those of the other two cities.

For firewood, although margins are highest in Mbeya and lowest in Dar es Salaam, the figures are slightly suspect. Transport is again the largest cost component in Dar es Salaam, while the producer margin dominates the cost structure in Mbeya and Shinyanga. In Mbeya, most traders are producers, so they capture the entire margin. This contrasts with the case in the other cities, where the trader margin is a small proportion of the final price of the firewood.

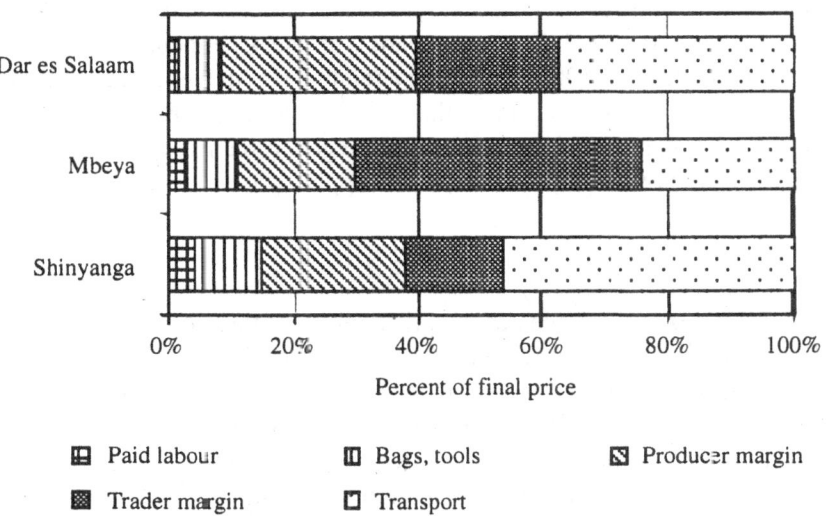

Figure 7.1 Cost structure for charcoal

112 *Woodfuel Markets in Developing Countries*

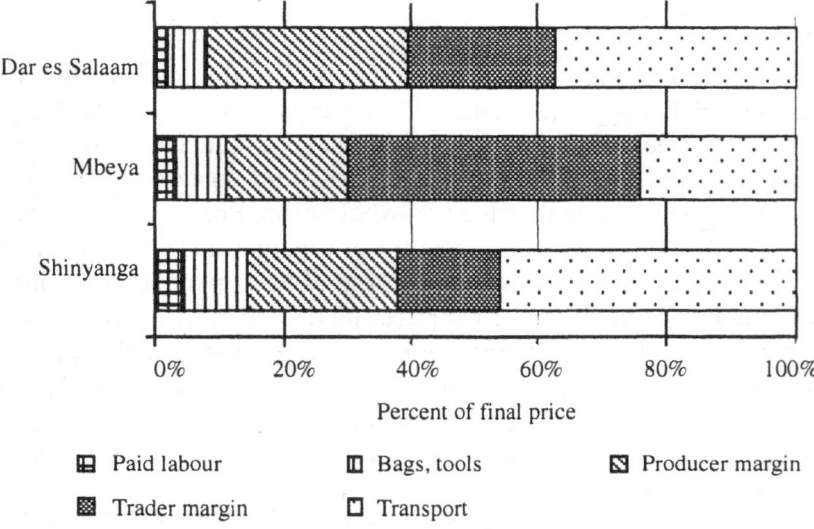

Figure 7.2   Cost structure for firewood

Taxes

Taxes and fees on the fuels, at both the rural and urban ends, total 7 to 8 per cent of the retail price for charcoal (per sack), and less than 1 per cent of the retail price for firewood. These compare with the 4 per cent of retail paid in taxes for woodfuel in 1986, indicating a relative rise in fees and taxes. A large percentage of these taxes is not collected, and woodfuel traders consider the taxes high. In many cases, they claim that the government reaps a larger margin than any of the other participants in the supply system.

Profitability

Table 7.2 gives a rough idea of the profitability of the woodfuel business for different participants. Charcoal producers have the best return on their labour in Shinyanga, as do wholesalers and, insignificantly, retailers in Dar es Salaam. Traders do much better than producers in general for charcoal on a per hour basis, though their per bag margins are sometimes lower. These margins are approximate and do not reflect costs such as capital, risk and, in some cases, equipment costs for traders. The return to labour for traders in Mbeya is unexpectedly low, despite the higher productive efficiency of the larger kilns burnt, and the potentially noncompetitive atmosphere. In fact, if the return to labour for the cooperative is considered separately, the margins would be larger. Additionally, the hours of labour per bag reported in Mbeya are much higher than that reported elsewhere, changing the return to labour considerably.

The urban minimum wage in Tanzania in 1990 was Tsh 2,750 ($US 9) per month. Clearly, the returns to labour obtained by workers in the woodfuel business are higher than this minimum wage. As a further comparison, an average government salary in Dar es Salaam was about Tsh 10,000 ($US 33) per month, or about Tsh 57 ($US 0.19) per hour. Again, urban participants in the woodfuel business do better than their counterparts in other employment. The considerable risk taken by wholesalers and the cost of capital investment are not included here. Certainly a government job includes benefits far beyond those offered by the hard life of a woodfuel seller. This comparison only serves to give some context to the earnings enjoyed by participants in this sector. Since rural earnings are generally lower than urban earnings, woodfuel producers also outperform many of their farming colleagues. To do so, they must work extremely hard using relatively labour intensive methods and spending sometimes long periods of time away from home.

For firewood, producers seem to do substantially better than traders. It is probable that the reported hours spent firewood harvesting are understated, particularly in Shinyanga.

Table 7.2 Return to labour

	Producer	Charcoal Wholesaler	Retailer	Firewood Producer	Trader
Dar es Salaam					
Margin Tsh/bag	188	132	187	2.0	0.2
Manhrs/ bag	4.14	0.68	1.50	0.57	0.18
Tsh/hr	45.4	194	125	3.5	1.1
Mbeya					
Margin Tsh/bag	110	233	317	5.6	1.8
Manhrs/ bag	5.5	3.1	3.0	0.06	0.25
Tsh/hr	20	75	105	93	7.2
Shinyanga					
Margin Tsh/bag	139	146	183	4.2	1.3
Manhrs/ bag	2.2	0.98	1.64	<.001	.01
Tsh/hr	63.4	149	112	4200	130

Intercity Prices and Variations

Though differences between the three cities in the study have been considered throughout this study, it is important at this point in the consideration of price efficiency to look explicitly at the spatial aspects of pricing. Tables 7.3–7.5 show the per kilogram prices for each level of the supply system, in each of the cities. Slight variations between the prices reported by different participants at the same level (e.g., end user purchase price and trader selling price) exist. However, these are mostly due to inexact matching of buyers and sellers and averaging errors. For the most part they are insignificant. There is a large gap only in the case of firewood, producer and trader purchase price. This is probably a result of inadequate knowledge of weights of fuel involved, as the trader often splits the wood into smaller pieces before selling. Unfortunately, the recording of the various sizes involved was incomplete.

Table 7.3 Fuel price build-up – Dar es Salaam

Tsh/kg (s)	1989 Dry	1989 Wet	1990 Dry
Charcoal			
Producer price	4.4 (2.0)	6.4 (2.6)	5.4 (1.9)
Trader purchase price	5.0 (3.0)	7.3 (4.2)	6.5 (3.7)
Trader selling price			
Bag	13.5 (1.9)	17.1 (2.5)	15.1 (1.9)
Kopo*	17.0 (4.7)	24.4 (6.4)	21.1 (5.5)
End user purchase price			
Bag	13.7 (2.9)	17.4 (3.2)	15.8 (2.5)
Kopo	17.1 (4.5)	24.4 (6.4)	20.8 (5.3)
% difference between bag and kopo	25%	40%	32%
Firewood			
Producer price	1.8 (2.2)	3.4 (4.3)	2.5 (2.9)
Trader purchase price	5.6 (23.0)	8.4 (33.9)	7.9 (33.4)
Trader selling price	5.2 (2.1)	7.4 (2.6)	6.6 (2.4)
End user purchase price	4.6 (1.5)	6.7 (2.0)	6.1 (2.4)

* A kopo is a commonly used measure for charcoal of about 1.5 to 3 kg, depending on city.

Table 7.4 Fuel price build-up – Mbeya

Tsh/kg (s)	1989 Dry	1989 Wet	1990 Dry
Charcoal			
Producer price	5.0 (3.3)	6.6 (4.5)	6.0 (4.2)
Trader purchase price	8.5 (5.2)	9.7 (6.2)	9.4 (5.7)
Trader selling price			
Bag	13.0 (1.4)	15.6 (1.5)	14.4 (1.5)
Kopo*	14.9 (11.1)	22.4 (6.9)	20.6 (9.4)
End user purchase price			
Bag	13.6 (2.0)	16.0 (1.7)	14.8 (2.1)
Kopo	17.2 (6.0)	23.9 (7.4)	21.7 (6.5)
% difference between bag and kopo	26%	49%	47%
Firewood			
Producer price	4.7 (2.6)	5.9 (3.6)	5.8 (3.1)
Trader purchase price	9.8 (5.4)	12.5 (6.7)	12.0 (6.7)
Trader selling price	8.1 (4.7)	11.0 (6.9)	10.7 (6.2)
End user purchase price	9.0 (6.0)	11.9 (6.0)	10.9 (5.8)

* A kopo is a commonly used measure for charcoal of about 1.5 to 3 kg, depending on city.

Table 7.5 Fuel price build-up – Shinyanga

Tsh/kg (s)	1989 Dry	1989 Wet	1990 Dry
Charcoal			
Producer price	3.3 (2.7)	4.0 (2.4)	4.0 (3.1)
Trader purchase price	3.9 (3.0)	5.5 (3.4)	4.6 (3.1)
Trader selling price			
Bag	12.7 (2.1)	17.1 (2.7)	15.6 (2.5)
Kopo*	18.9 (3.9)	27.9 (7.0)	24.4 (4.7)
End user purchase price			
Bag	12.9 (1.8)	17.3 (3.2)	16.9 (2.9)
Kopo	18.1 (6.0)	25.4 (8.1)	25.2 (7.7)
% difference between bag and kopo	40%	47%	49%
Firewood			
Producer price	0.45 (0.43)	4.15 (8.5)	4.18 (8.5)
Trader purchase price	0.87 (0.40)	1.5 (0.83)	1.2 (0.51)
Trader selling price	4.2 (1.8)	8.6 (5.8)	8.3 (5.8)
End user purchase price	4.5 (3.2)	9.1 (9.4)	8.8 (6.8)

* A kopo is a commonly used measure for charcoal of about 1.5 to 3 kg, depending on city.

Price Differences Between Bulk and Small Purchases

A point of interest is the difference between per kilogram prices of large (bag) and small (kopo) purchases of charcoal. As shown in Tables 7.3, 7.4, and 7.5, the differential ranges from 25 per cent to 50 per cent, and has increased in the past year. The per kilogram kopo price is rising faster than the per kilogram bag price. Nevertheless, the kopo is the unit in which most consumers buy. Because of the value added, the price is increased by breaking the bag into smaller measures as nearly 15–20 per cent of the product is lost in the fines (charcoal dust). The small volume of sales for many retailers and their need to make a living selling charcoal also contributes to the price increases. This price differential is even greater in the wet season. Since it is the mass of consumers, and the poorest, who pay the higher prices, the poor pay for the convenience of being able to pay on a day to day basis. However, a chronic lack of cash makes convenience a necessity for most users. These seasonal and annual changes will be discussed further in the following sections.

High Mbeya Firewood Prices

Firewood prices are the highest in Mbeya, even though that is the one town of the three that appears to have plentiful firewood supplies. About 23 per cent of households in Mbeya purchase firewood for use even though it is more expensive than charcoal on a calorific basis. There are no equipment costs, which might encourage the lowest income users to remain with firewood as their fuel of choice. The amount sold to each customer is small, which implies both that the levels are appropriate for low-income customers, and that many customers supplement their purchases with gathered fuel. For consumers that gather part of their firewood themselves, it makes sense to supplement that supply with purchased firewood, rather than to involve themselves in a different fuel such as charcoal that cannot be burnt in the same stove or mixed with the firewood. This may be true even if they are ultimately paying more for the fuel on a calorific basis. Although a third of households in Mbeya use both charcoal and firewood, many of these change from firewood to charcoal during wet season, when firewood is harder to dry.

Trader marginality Another reason for high firewood prices in Mbeya may be the marginality of most if its firewood traders. The total amount offered for sale by the majority of the traders is very small, consisting only of the amount the seller could carry from the timber plantation from which they collected their supply. In order to make a living in this situation, the selling price needs to be higher.

Influence of Transport Distances and Road Condition

At the producer level, charcoal prices are lowest in Shinyanga. For the end user the purchase price is the highest. This reflects higher transport costs and distances in the region. Figure 7.3 illustrates the influence of transport distances and road condition on the producer, or ex-kiln, price for the three cities. As expected, there is a discernible ex-kiln price decrease as distance increases and road condition worsens. This effect is slightly weaker in the dry season.

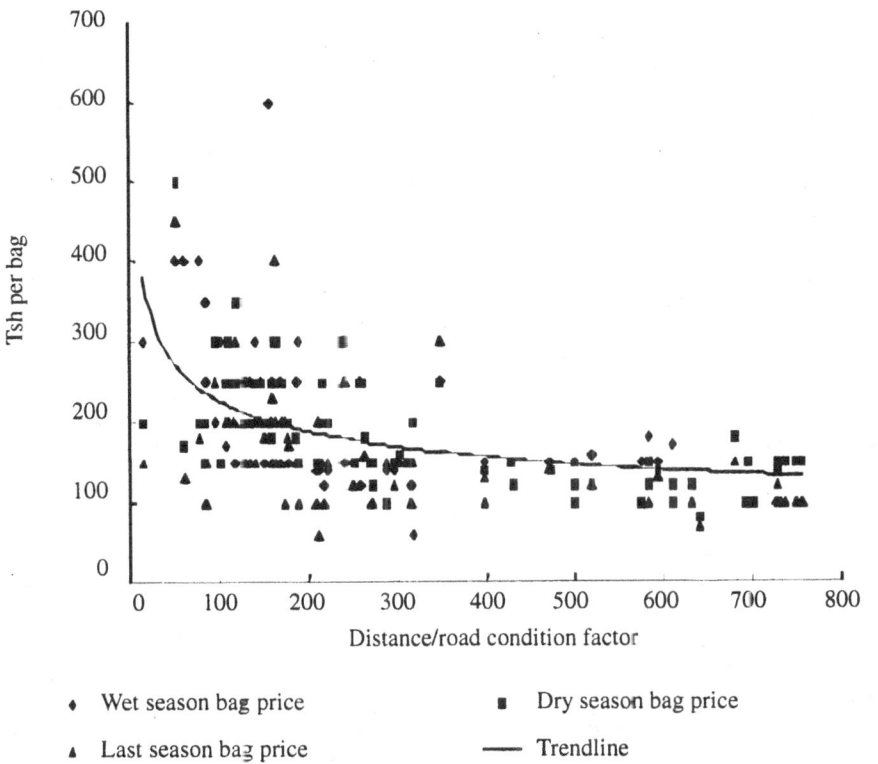

Note: higher factor numbers indicate longer distances and/or worse road conditions.

Figure 7.3 Relationship between kiln price and distance to town and road condition

Since transport prices are a large percentage of the final price of the fuel, distance from the market should negatively influence the ex-kiln price of the fuel, as is the case in Shinyanga. This contributes to the low relative margins of the wholesalers in that city. In Mbeya, though the producer prices are the highest of the three cities, the end user purchase price is the lowest, reflecting the lower transport charges and taxes and fees paid in the region.

Table 7.6 shows the correlation between distances travelled and ex-kiln prices. There we can see that when the total distance travelled by the charcoal from the kiln to the end user is considered, there is a difference between cities. In Dar es Salaam, ex-kiln price is negatively related to both total distance and dry season road condition. This is the result that would be expected, as longer distances travelled on poorer roads engender lower prices at the production site.

Table 7.6   Relationship between road distance and producer price

Correlation	Total distance (Season)		Side road (Season)		Road condition (Season)	
	Dry	Wet	Dry	Wet	Dry	Wet
Charcoal	-0.57 **	-0.54 **	0.001	0.025	-0.61 **	-0.55 *
Dar es Salaam	-0.42 **	-0.41 **	0.11	0.03	-0.35 *	-0.23
Mbeya	0.51	0.34	0.47	0.65 *	-0.48	-0.32
Shinyanga	0.53 **	0.62 **	-0.37	-0.07	0.70 **	0.23
Firewood	0.91 **	0.88 **	0.49	0.53	0.90 *	0.79

*   indicates significance at the 95% level.
**   indicates significance at the 99% level.

In Mbeya, there is a negative correlation with road condition only, indicating that this is the factor particularly driving down ex-kiln prices. Total distance is positively related to price. This may reflect the differences between producers in different areas or working with different wholesalers. Those working for large wholesalers may get a lower price (supplemented by food and materials) for their fuel than those working independently or for small wholesalers, regardless of where they are set up to produce. This may be true particularly since the licensing system mandates that the production be done in specific areas chosen by the forestry department. The wholesalers then deposit the producers working for them in the designated area and set them to work at standardized terms. Therefore, the effect of distance would not be reflected in the ex-kiln price in this case.

In Shinyanga both the road condition and total distance are positively related to price, meaning that ex-kiln prices are higher in more distant places regardless of the length and difficulty of the transport involved. This result is probably dominated by results from the supply area for Shinyanga fuel which coincides with that for two other larger cities, Mwanza and Tabora. Therefore, though the distances to Shinyanga from these locations may be greater, the competition from the other towns keeps the ex-kiln prices high. In fact, harvest areas further from Shinyanga are often closer to one of the other cities, and so the closest city may drive the kiln price. Additionally, the level of scarcity of the fuel in the three competing towns will influence the final price of the fuel, and so indirectly the ex-kiln price. If suppliers for the town with the largest supply problems have trouble finding fuel, they should begin to increase their price offers to producers to ensure that they have access to a regular supply. It is not possible to determine which or in what combination these forces are working given the information at hand. However, a further investigation could prove interesting.

Distance from Main Road

When looking at the relationship between the ex-kiln price and the distance of the harvest area from the main road ('side road' distance in the table) again it is expected that the further the distance a harvest area is from the main road, the lower the producer price will be. For the buyer, this part of the trip is the most difficult in terms of wear and tear on buyer and vehicle, time spent, and road difficulties encountered. However, in this case the correlation is low and the sign is wrong, except for the case of Shinyanga. Therefore, in most cases the total distance the supply area is from the city and the road condition are more important to the ex-kiln price than the distance off the main road. This may indicate that buyers would rather drive further on the main road to another supply area than delving deeper into the existing supply areas *via* extremely poor side roads. In their view, it may be that an extra 20 kilometres on a good road is preferable to an extra five kilometres on a side road.

However, only in Mbeya is the relationship between the ex-kiln price and the distance travelled on side roads strong. The relationship between licensing, wholesalers, and producers may again be at work here to confound such relationships. As well, some producers indicated that they moved to different production areas in the wet season. The producers indicated that these areas have a different soil type, different tree species, and less rain, making charcoal production in the rainy season somewhat easier. These wet season production areas may be further off the main roads.

Firewood

For firewood, all relationships are positive, and both the total distance and road condition have strong positive correlation with the producer price, particularly in the dry season. This means that producer price increases as distance from the city increases and road conditions worsen. This relationship is driven by the situation in Mbeya, where firewood from more distant places means it was harvested from natural forests rather than private trees and government plantations. These trees fetch a higher price for the producer, and therefore the relationship is skewed in the positive direction.

Integration of Markets over Space

Most of the literature exploring the organization and competition of food markets in Africa attempts to link the prices of various markets through some mechanism, often correlation estimates, in order to determine whether the intermarket price differences are explained by arbitrage and transportation costs, indicating a competitive market (cf. Jones 1972, Raju and Bhatt 1985, Southworth et al. 1979, Saul 1987). The use of correlation coefficients in this way has been severely criticized (Harriss 1979) as being spurious. In any case, such an analysis requires historic prices from several towns that are net suppliers to the larger population areas. In the woodfuel sector, fuel supply is generally very localized and the fuels, charcoal in particular, are generally not marketed or even used in the areas which supply them. Additionally, historic fuel price data is difficult to come by even in population centres in Tanzania, and nonexistent in smaller towns. Therefore, with such a different situation and different data, these tests are not expected to reveal much. In fact, when a correlation test of spatial arbitrage was attempted using historical retail prices on the cities in this study, low coefficients were expected and obtained.[2] The one relatively high correlation, that between Dar es Salaam and Mbeya, might be partly the result of information carried between the towns by the truck traffic that is common between them. However, it is just as likely to be unrelated to any interaction between the cities, because little woodfuel actually moves from one town to the other in any case. The price differential between the towns is not wide enough to encourage spatial arbitrage at such levels, especially since the transport cost of the fuel already occupies a large proportion of its final price.

Overlapping Markets

In this case, the question of overlapping hinterlands or fuelsheds is interesting. There are overlapping supply areas in the western supply region of Dar es Salaam, where supply is divided between it and Morogoro, a good-sized regional capital. Because of regional supply difficulties, the supply area for Shinyanga overlaps that for both Tabora (a regional capital) and Mwanza (the second largest city in the country). It would be interesting to examine the forces that determine to which city supplies go. Unfortunately, data is not available to do so. However, an examination of prices in towns with overlapping supply areas, along with inquiry into personal bonds and other linkages which impact this determination would prove fruitful areas for further research.

Spatial Price Differences within Dar es Salaam

A last topic in the section is the differences between prices in different areas in the same city, Dar es Salaam. In Table 7.7 some of the fuel price data for the city has been split into three large geographic districts. Though each of the districts is mixed, Kinondoni is a more urban, higher income and more residential district in general. Temeke is the most industrial district, and is less densely populated than the other districts. Ilala is a fairly dense and more commercial district, although it contains a fair amount of densely residential population.

We can make some observations about the price differentials in the different districts. Charcoal prices in Temeke, which has been characterized as a more industrial district, are significantly lower than in the other districts. Although this is particularly true for sales by the bag, sales in small amounts were priced similarly to Ilala. Kinondoni has the highest kopo selling prices, probably due to the higher demand of this residential district, and influenced by the more affluent tenor of some of its residents.

The average price at which sellers of all types buy bags of charcoal is much lower in Temeke than in the other districts. Temeke is located close to the longest-running fuelwood supply zone in the Dar es Salaam area. Traders and producers mentioned that it was sometimes worth leaving their fuel in one of the peri-urban markets, despite the lower price they got for their product, because it saved going into the centre of the city, thereby saving time, transport costs, and, sometimes, fuel taxes on the fuel. However, there is no clear potential for arbitrage, since the prices per bag in Temeke are Tsh 40 to Tsh 50

Table 7.7   Fuel pricing in Dar es Salaam districts

	Season	Kinondoni	Ilala	Temeke
Charcoal (tsh)				
Bag selling price	Dry	671.2	682.1	630.5
	Wet	747.0	774.1	670.0
Kopo selling price	Dry	35.8	31.4	30.4
	Wet	40.0	37.8	35.8
Bag buying price	Dry	268.4	265.0	160.0
	Wet	388.0	417.0	314.8
Sellers' margin	Dry	148.4	166.4	55.0
	Wet	164.3	189.0	23.4
Daily sales of bags	Dry	7.1	4.8	3.7
	Wet	9.3	7.5	4.5
Daily sales of kopos	Dry	18.1	13.8	15.9
	Wet	19.3	19.7	17.9
Bags sold per year		3054	2513	1520
Selling price per kg	Dry	5.82	8.50	5.48
	Wet	6.79	9.38	6.41

lower than bag prices in the other districts, while transport costs between districts average around Tsh 65 per bag. It seems that sellers in Temeke are not able to capitalize on the low wholesale buying prices of the fuel there, since average sellers' margins in Temeke are markedly lower than those in the other districts in the table. It is not clear why their margins are so low. Perhaps it is because they are smaller scale than sellers in other districts. This may contribute to their smaller margins.

Firewood prices are lowest in Temeke while those in Ilala are significantly higher. Perhaps the demand in Temeke, as a partly industrial district, is relatively low, and users are able to collect some of their fuel on their own. Again, the proximity of Temeke to the supply areas might also help explain the lower prices. The high prices in Ilala district may be contributed to by shortages caused by a reduced number of firewood sellers, as most people in that district use charcoal or other fuels. More data needs to be obtained in order to sift out the reasons for these differences.

Temporal Issues

Seasonal Prices

There are significant price changes for woodfuels between the dry season, generally considered to be from about April to October, and the wet season, stretching from November to March. Most roads in Tanzania, and particularly the secondary and tertiary roads that service woodfuel-harvesting areas, are dirt or gravel. Because such roads are significantly worse in the wet season, prices are expected to rise in the wet season due to transport cost rises. The producers of woodfuels, primarily farmers, are busy at their agricultural activities in the wet season. They have less time to produce, causing a drop in supply and an increase in the opportunity cost of their time. Those that do produce, then, need a higher price to compensate.

These expectations are reflected in the data. Table 7.8 shows the seasonal price changes in each town. In Dar es Salaam and Mbeya the rise in charcoal producer prices is not reflected completely in end user prices when the fuel is bought in bulk, but there is a similar price rise when small quantities are sold. In Shinyanga end user purchase prices reflect more than the rises in producer prices during the rainy season. Since a higher percentage of charcoal producers work full-time in Mbeya and Shinyanga, wet season supply disruptions are less common, explaining the lower seasonal differential in producer prices as compared to Dar es Salaam.

In Dar es Salaam and Mbeya, wholesale bag prices do not jump nearly as high in the wet season as do producer prices. Since wholesalers absorb increases in transport costs, it seems that their profits would be decreased during the wet season. In Mbeya and Shinyanga, wholesalers increase their prices to reflect both the producer price changes and the transport charges, which rise during the rains to 11 per cent to 30 per cent over dry season charges. Wholesaler margins increase significantly in Mbeya and Shinyanga but remain at dry season levels in Dar es Salaam.

This evidence supports the idea that wholesalers in Dar es Salaam do not control the competitive market there, but that wholesalers in the other two cities have some liberty to reap excess profits, particularly in times of limited supply and increased demand like the wet season. From an earlier discussion, only medium-sized wholesalers have margin increases in the wet season, and most of the largest wholesalers in Shinyanga and, to a lesser extent, Mbeya are considered 'medium' in our scaling. Again, it is the wholesalers in Shinyanga and Mbeya who have wet season margin increases.

Table 7.8 Seasonal change in woodfuel price 1989–90

	Dar es Salaam	Mbeya	Shinyanga
Charcoal			
Seasonal change (dry to wet)			
Producer price – bag	45%	32%	21%
Wholesale price – bag	27%	18%	34%
End user price – kopo	43%	39%	40%
Margin change (dry to wet)			
Wholesaler	0%	+19%	+14%
Retailer	+2%	-18%	0%
Firewood			
Seasonal change (dry to wet)			
Producer price	89%	26%	38%
End user price	46%	32%	96%

*   The change in the consumer price index was +19%.

While retailers in Dar es Salaam do slightly better than the wholesalers in the wet season, with their margins rising by 2 per cent, retailers in Shinyanga maintain dry season margins. In Mbeya, retail margins drop sharply. This takes place despite the fact that wholesale prices do not increase as much as retail prices, and may be attributable to the number of wholesalers who double as retailers, and the number of retailers who get their own fuel.

Firewood

The seasonal price changes for firewood show similar trends. In Dar es Salaam, the end user prices do not come close to reflecting the producer price increases. A common complaint by Dar es Salaam firewood traders was that 'no one' uses firewood in the wet season, because the fuel is wet and therefore difficult to light. This probably explains the relatively low end user purchase prices in the rainy season. In Mbeya and particularly Shinyanga the seasonal fuel switching does not take place to the same extent, and so the scarcity of firewood caused by the decrease in producer activity results in a price rise. In Shinyanga, this is exacerbated by the fact that most buyers of firewood are informal-sector businesses for whom there is no substitute fuel.

Storage

There is a potential for storage to mediate these seasonal price variations. Most traders who do not store made that decision based on lack of capital resources, because they need to sell their stock as fast as possible to get money to live on and with which to buy new stock.

A very few producers store fuel that they have harvested or produced in the dry season in order to sell at higher prices in the wet season. Most producers did not, and many thought that the prices they received were 'fair' and needed the cash the fuel provided in the off season for agriculture. Others were working on another's license and had no control over their production, per se. Additionally, the capital risk is high, because the fuel can be destroyed or stolen and prices can change from expectations. In any case the differences in prices are relatively small, and risk aversion is to be expected from a group that is generally cash poor.

Seasonal Index

The potential importance of speculative storage and the determination of whether it is a practice which should be facilitated in order to improve the overall efficiency of the system, can be tested by calculating seasonal indexes of historic prices of the fuels in the cities under study. Table 7.9 shows the seasonal indexes of retail fuel prices in the three towns for the period 1977–90. The wet season takes place during the first and fourth quarters, and the dry season spans the second and third quarters. Because there is only quarterly data, into which the seasons do not exactly fit, and because that data is spotty and contains outliers, these tests give only a first feel for the trends shown by the prices. They show a moderate 4 per cent to 13 per cent seasonal price rise between dry and wet seasons for charcoal. A wet season price *drop* in Dar es Salaam was found, and Shinyanga showed a slightly different but extremely moderate seasonal price change cycle. There were very few data points for firewood in any of the towns, so results were unable to be calculated.

Reasons not to store It seems clear from these results why there is little speculative storage of charcoal in any of the cities. The relative returns, in the long run, are not particularly great, and the variability of the real rises in price, as inferred from the standard deviations, are great, making the risk of storage greater. Adding that to the extra cost in time and labour of storage, along with the risk of theft and the smaller risk of spoilage, and the opportunity is not particularly appealing to most sellers.

Table 7.9   Index of quarterly retail prices of woodfuel 1977–90

Index (sd)	1st quarter (Jan.–March)	2nd quarter (April–June)	3rd quarter (July–Sept.)	4th quarter (Oct.–Dec.)	Range
Charcoal					
Dar es Salaam	88 (21)	105 (33)	99 (26)	87 (24)	18
Mbeya	116 (38)	106 (39)	109 (28)	120 (45)	14
Shinyanga	103 (40)	104 (32)	108 (23)	107 (18)	4

Source: from data obtained from the Tanzanian Statistical Office. The index is the average quarterly percentage of the centred one-year moving average.

Even more importantly, the price of credit and the rate of inflation both influence the decision of whether to store fuel. If a seller is borrowing money to buy extra fuel to store, a credit price which is higher than the added profits will make storage unprofitable. In Tanzania, only small sellers admit to much borrowing, but even for large traders, if credit prices are high, it may be a better business opportunity to lend money or invest it in another venture than to store fuel. Inflation can quickly delete any profits in a long term investment if the return on the investment is lower than the rate of inflation. If the rate of inflation is not known, there is a large risk associated with it.

Additionally, in Mbeya and Shinyanga the wholesaler margins already rise in the wet season, giving little incentive for storage. If data for wholesale prices were available, they might tell a different story, but the evidence available points strongly away from the viability of seasonal arbitrage.

Relationship between Historic and Seasonal Price Changes

To further explore this result, the historical retail price data was used in a regression analysis to find the relationships in the three cities studied between the historic price changes for charcoal, the seasonal variations in charcoal prices, and the historic changes in the price of kerosene. The results of the analysis are shown in Table 7.10. Dummy variables were used for seasons or quarters, and outliers were removed from the Dar es Salaam and Shinyanga data. A dummy variable was used for Mbeya data for the years 1983 and 1983. There were a series of economic disruptions in those years that led to uncharacteristically high prices for woodfuels during that time. Following that period, prices returned to near what they had been before the disturbance. The dummy variable ensures that the results are not inordinately influenced by the anomalous period. In the all cities case, city dummy variables were introduced to determine if there were differences between the cities. Both city dummy variables were significant, indicating that there is a discernible difference between the cities.

Table 7.10   Factors influencing charcoal prices

B(t)	Time	Season	Kerosene price	1982–83	Mbeya dummy	Shinyanga dummy	Adj. R^2
Dar es Salaam	0.0015	-0.053	0.094 *	–	–	–	0.06
	(1.05)	(-0.95)	(1.63)				
Mbeya	0.024 **	0.072	-10.83	1.30 **	–	–	0.58
	(4.90)	(0.43)	(-0.66)	(4.76)			
Shin-yanga	0.034 **	0.025	37.94 **	–	–	–	0.72
	(7.89)	(0.21)	(2.17)				
All cities	0.011 **	-0.082	0.018	0.81 **	0.51 **	0.55 **	0.44
	(4.13)	(-0.91)	(0.13)	(6.26)	(2.05)	(2.17)	

*   indicates significance at the 90% level.
** indicates significance at the 95% level.

Source: Tanzania Bureau of Statistics.

The results reinforce those achieved when looking at the seasonal indexes. The seasonal coefficients in the regression are not significant for any of the cities, and it seems that there is little price difference due to seasonal influences. As the seasonal indexes brought out, what little seasonal influence there is on price turns out to be positive for Dar es Salaam and Shinyanga, rather than the expected negative influence. Again, this supports the general lack of storage in the three cities, and suggests that the decision not to store has been a rational one over time. The results of the regression analysis will be discussed further in the next section.

Historical Prices

The data and analysis seen in the previous section contains much information about the historical trends of prices in the three cities. The price data for charcoal for the past 15 years is contained in Figure 7.4. It depicts both the

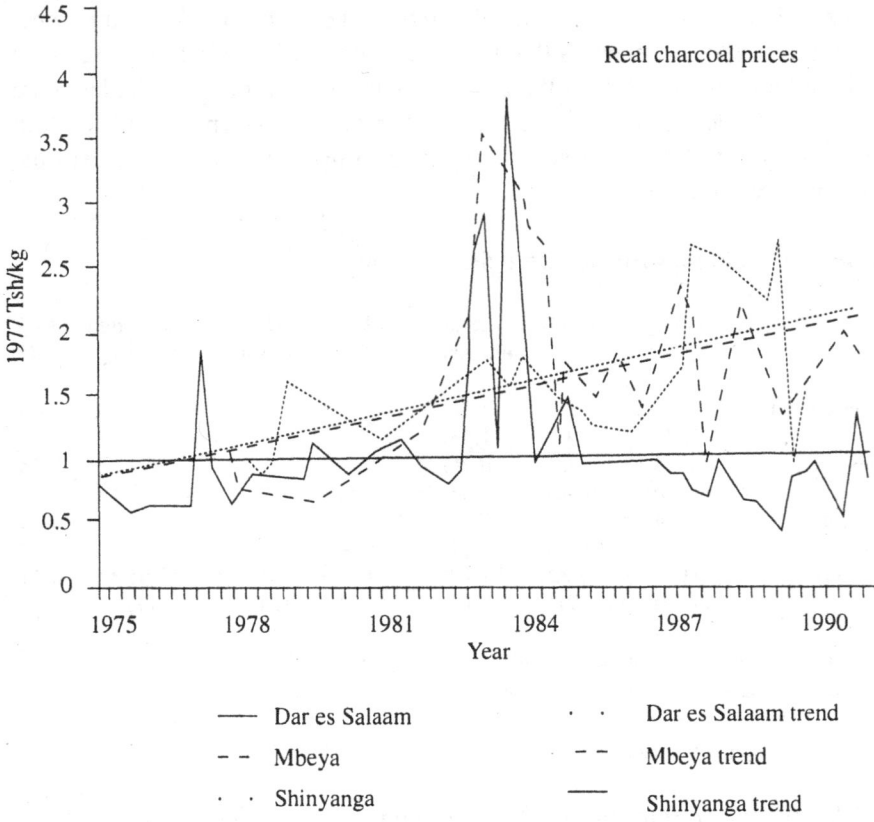

Figure 7.4 Real charcoal prices and trends

real price of charcoal and the trend (calculated from the regression analysis reported in Table 7.10) of the real charcoal price in each town. There has been no definite rise in real prices in Dar es Salaam, but there has been a discernible increase in both Mbeya and Shinyanga, with the growth in Shinyanga being greater. This is reflected in the regression results in Table 7.10, where the variable for time is significant for both Mbeya and Shinyanga.

Reasons for Price Rises

Scarcity The potential reasons for these differences are many. The first is the possibility that the increasing prices in Mbeya and Shinyanga reflect increasing scarcity in those areas. Table 7.11 displays the perceptions of producers in the supply areas for the three cities regarding the supply of woodfuels. Far more producers perceive increasing scarcity of harvesting opportunities in Mbeya and Shinyanga than in Dar es Salaam.

Certainly, there is a dearth of biomass near to Shinyanga, and the distance charcoal travels to get to town reflects its relative scarcity. An additional constraint on charcoal supplies for Shinyanga is that it shares parts of its supply areas with Mwanza, where there are also inklings of scarcity, and Tabora. Forests around Tabora supply all three towns. In Mbeya, the distances travelled by the fuel are also said to be increasing, and the population in each of the towns is increasing. Although it is possible that supply is not keeping up with the demand, causing price rises, in one supply area for Shinyanga some full-time producers complained that they were having trouble selling their fuel. Some are even switching back to agriculture due to low producer prices and the hard work involved in charcoal making. This implies that on the outskirts of the supply envelope for Shinyanga demand, and prices, are not yet quite high enough to spur their participation. Conversely, in Dar es Salaam, producers said there was never any trouble selling their fuel.

Constraints to competition Additionally, these indications point towards a constraint to competition in the wholesalers of Shinyanga and Mbeya. Wholesaler control over prices could cause price rises such as those seen in the historical record, and if so would represent a great inefficiency in the system, and a welfare loss by consumers.

Single year price variation A look at producer, wholesale and retail prices over a shorter period may lend insights that retail prices over a long period could not. Table 7.12 displays this more detailed data. Its limitations stem from the fact that it spans only a single year. As with seasonal price changes, real annual price changes show a greater gain for the retail price than for the wholesale price. This is particularly true, and most surprising, in Mbeya where the structure of the market has indicated that the wholesalers might be expected to have enough control over the market to force real price rises rather than losses. Again, the fact that this is a one year snap shot reduces its usefulness – there are many explanations for the decrease in wholesale price. In Shinyanga, the boom in gold has raised the population and incomes there, possibly at a rate greater than that in other parts of the country. This may affect the calculation of the real rise in prices. However, it may not be enough to explain the large wholesale and retail price increases there. The producer price rises were negligible in the Shinyanga supply area, even with the assumedly greater demand from the various cities which draw supplies from the area. These trends again give an indication of wholesaler control in that city, and reinforces the indications seen before, including those in the historical data.

Table 7.11 Percent of participants agreeing with supply scarcity questions

% agreeing	Dar es Salaam	Mbeya	Shinyanga
Producers			
Are trees in this area getting more scarce?	53.5	70.2	61.8
Do you harvest in different areas than you did 5 years ago?	32.3	63.6	75.0
Sellers			
Is it harder to get fuel supplies than it used to be?	71.1	61.9	56.1

Producers in Dar es Salaam have the largest real price rise, strengthening the argument that the individualistic and independent village-based production of Dar es Salaam allows producers more control over prices. It is possible also, however, that factors such as better information and increased demand are responsible for the favourable price turn for the Dar es Salaam producers.

Although the firewood price data are suspect, they display a trend that shows prices to be clearly rising fastest in Shinyanga. The prices are barely rising in real terms in Mbeya, which is expected due to the supply structure there, and are rising fairly quickly in Dar es Salaam. Shinyanga's meteoric price rises are attributable to the lack of fuel flexibility by those who use firewood. Many of the users who purchase the fuel are beer brewers, and there is no easy substitute for large pieces of firewood in that business. As more people use charcoal and other fuels, and as others continue to collect firewood from their farms for their personal use, firewood prices may continue to rise. The price rise may also be due partly to supply constraints, since firewood costs more to transport, and is generally collected from closer, more heavily utilized areas. We saw that the distances travelled by firewood in Shinyanga are already the largest of the three cities, and it is possible that these distances are getting greater, increasing the price of firewood and decreasing producer prices.

Substitution The degree of substitutability of woodfuel with other fuels in the respective cities is a factor which influences real price increases. In Shinyanga

Table 7.12 Annual changes in woodfuel prices 1989–90

	Dar es Salaam		Mbeya		Shinyanga	
	nominal	real	nominal	real	nominal	real
Charcoal						
Annual change (1989 to 1990)						
producer price – bag	23%	+4%	20%	+1%	21%	+2%
wholesale price – bag	15%	-4%	9%	-10%	31%	+12%
end user price – kopo	22%	+3%	26%	+7%	39%	+20%
Firewood						
Annual change (1989 to 1990)						
producer price	39%	+20%	23%	+4%	72%	+53%
end user price	33%	+14%	21%	+2%	102%	+83%

* The change in the consumer price index was +19%.

and Mbeya firewood provides stiff competition to charcoal, but there is little competition from 'modern' fuels such as electricity and LPG, or even kerosene for cooking fuel (the primary use of woodfuels in these areas). Woodfuel demand, particularly demand for charcoal among the poorest people, is considered to be highly sensitive to price variations, and some people use both charcoal and firewood for different applications (cf. Stevenson 1990, Openshaw and Feinstein 1989). However, once having switched to using charcoal (including buying the stove it requires) most people appear to be reluctant to switch back to firewood unless the price of charcoal rises significantly in relation to the price of firewood. Therefore, given this reluctance to change away from charcoal, traders in Mbeya and Shinyanga have some leeway to raise charcoal prices.

In Dar es Salaam, the availability of alternate fuel sources works to dampen woodfuel prices, though there are various non-price constraints to fuel switching which also figure in fuel choices. Kerosene and electricity prices have been controlled and real prices have been stable or falling until very recently (Hosier and Kipondya 1993). This could have a negative effect on woodfuel use and pricing.

To emphasize this, Figure 7.5 and Figure 7.6 show price trends for alternative fuels. Though not directly comparable, the graphs show the direction the price is going compared with that of charcoal. Figure 7.5 reveals that kerosene prices have been stable in real terms in all three cities until the last two years, when real prices have doubled. This might have had an effect on woodfuel use in the last few years, and even could have influenced the prices of those fuels as they followed its price rise. The regression results in Table 7.8 show a positive and significant relationship between charcoal prices and kerosene prices in Dar es Salaam and Shinyanga, suggesting that the prices of the two fuels do follow each other. Mbeya does not show the same tendency. As seen in Figure 7.5, in years which saw real price rises for charcoal in Mbeya and Shinyanga and stable real prices for the fuel in Dar es Salaam, it would seem that the influence of kerosene would have been to encourage switching from woodfuels to kerosene. A detailed discussion of fuel switching is beyond the scope of this paper, but suffice it to say that one of the factors consumers consider before switching is the reliability of the supply of a fuel. Kerosene supply, as with other commercial fuels, has not been consistent in the two smaller cities, and even in Dar es Salaam the supply of kerosene at

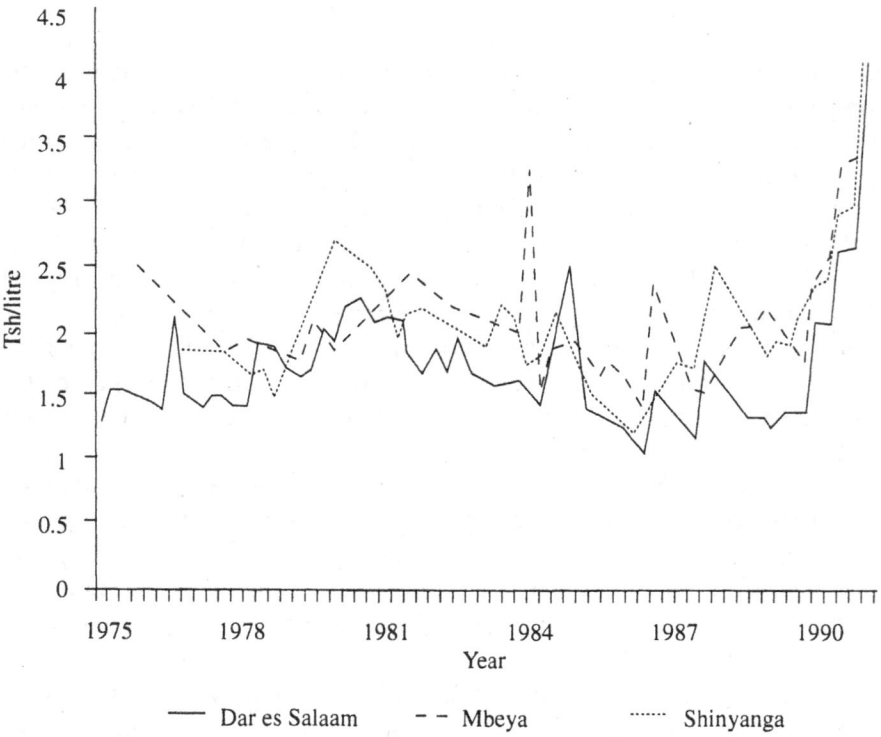

Source: Tanzania Bureau of Statistics.

Figure 7.5 Real kerosene price

official, controlled prices has always been inadequate. This, and the supply interruptions, had a discouraging effect on potential consumers, particularly in Mbeya and Shinyanga. Therefore, its price could not play the stabilizing role it might have in Dar es Salaam, or in the case of a free market for commercial fuels.[3]

Figure 7.6 graphs electricity prices in Dar es Salaam for the same time period. The real price figures show that prices have been declining steadily during the time period, though the fall has slowed, or perhaps stopped (with new tariff increases in January 1991) in the last couple of years. Electricity is an alternative to charcoal mostly in Dar es Salaam, and it might have influenced the steadiness of the real price of charcoal in that city. Although low income users are dissuaded from using electricity by the cost of hook up and appliances,

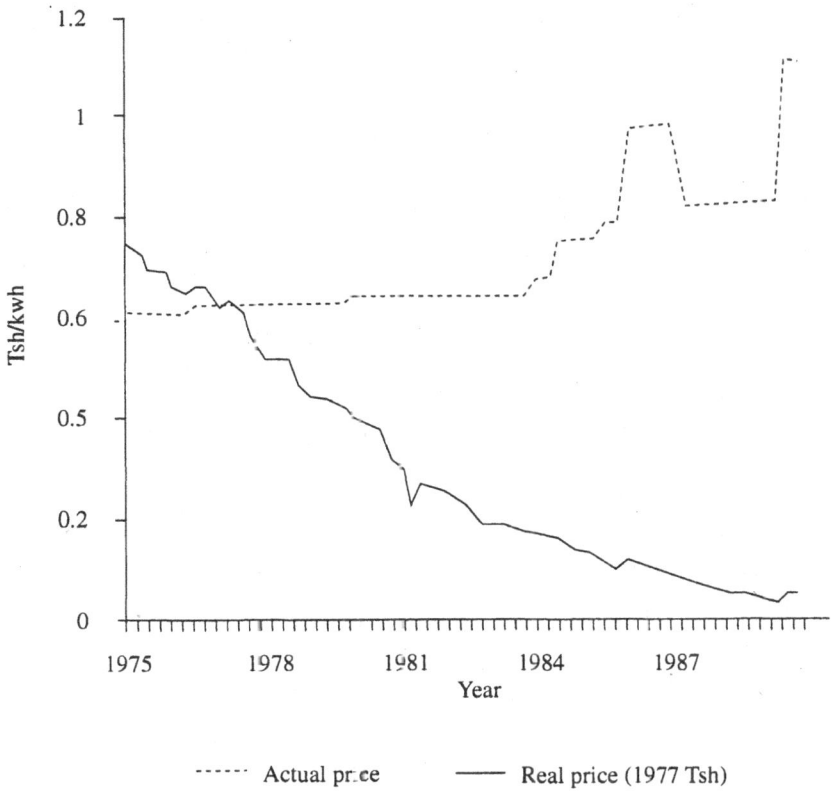

------ Actual price   ——— Real price (1977 Tsh)

Source: Tanzania Bureau of Statistics.

Figure 7.6 Electricity prices in Dar es Salaam

middle and high income users are many and there the competition with woodfuels is more direct.

Even where real prices have not increased, falling incomes in the past several years mean that the purchasing power of consumers has decreased, and that energy is occupying a larger percentage of most people's monthly budgets.

In this chapter, the effects of performance on the competitiveness and efficiency of the woodfuel market system have been further explored. It seems clear that prices are rising most quickly in Shinyanga, though in Mbeya wholesalers are taking home the largest margins. Because transport costs are found to profoundly affect the prices of the fuels, spatial arbitrage is not viable for wholesalers. Seasonal arbitrage was also found to be unattractive, since prices did not change enough in the long run to be worth the investment in the short run, and because wet season margins rose for wholesalers in any case.

Summary

Producers were shown to do best in the independent atmosphere prevailing in the Dar es Salaam supply areas, and wet season producers did especially well. Retailers also had increases in prices over the year studied, but their seasonal margins were fairly static and were reduced in Mbeya.

Firewood prices went up greatly for producers in Dar es Salaam and retailers in Shinyanga, although the margins were highest for traders in Mbeya and lowest in Dar es Salaam. It indicates that the traders in Shinyanga may soon catch up to the margins of their colleagues in Mbeya, though their costs will continue to increase while those in Mbeya will stay fairly static.

The next chapter will synthesize the results of this and the preceding chapters and make more sweeping conclusions about the workings of the woodfuel markets in the three cities studied.

Notes

1   A measure of charcoal weighing between 1.5 and 3 kilograms, depending on the city, which is generally sufficient for a day's cooking needs.

2 | | | MBY | SHY |
    | Charcoal | DSM | 0.53 | -0.1 |
    | | MBY | – | 0.30 |
    | Firewood | DSM | 0.23 | 0.02 |
    | | MBY | – | 0.14 |

   Retail prices from 1975-1989 were used to calculate simple bivariate coefficients of correlation between city pairs.
3 For more detail on this subject, see Hosier and Kipondya (1993).

8 Policy Implications and Conclusions

The structure and function of the woodfuel system in Tanzania has been discussed throughout this book. The research has attempted to verify or dispel widespread suppositions that the private market system that characterizes the delivery system for woodfuel in Tanzania and other developing countries is noncompetitive and inefficient. It also refutes the idea that there are temporal and spatial price vagaries that may be associated with collusive and monopolistic elements in the trade. These assumptions extend also into other informal and formal market and delivery systems in developing countries. The follow-up to such assumptions is that the particular system in question is in need of intervention and 'fixing'. The validity or error of these assumptions is important because they define the tenor of the policy decisions that follow from them, and hence the health and development of the woodfuel (or other) delivery system. This in turn impacts directly on the consumers of woodfuels, and affects the allocation of resources both within that sector and in the economy as a whole. The sustainability of the woodfuel system as major source of energy for the country and the sustainability of the woodlands that supply the sector are in turn impacted.

Competitiveness

The discussion points to several findings. The most significant is that the woodfuel market in Dar es Salaam gives every indication of operating as a competitive market. This is due to:

- a competitive structure;
- large numbers of small-scale producers, wholesalers and retailers;
- competitive conduct including open access to the market and freedom from collusive practices;
- efficient performance, including competitive pricing and margins for

Policy Implications and Conclusions 137

participants at different levels of the system, and moderate seasonal and historical price fluctuations.

The conclusion is not so facile for Mbeya and Shinyanga, because of the large and perhaps controlling wholesalers and limited access to licenses, but even there the evidence is mixed. For firewood, a limited demand, and therefore a small market, also impedes the smooth workings and competitiveness of the markets in Shinyanga and Dar es Salaam.

Structure of Supply System

Seller Concentration

In general, the structure of the woodfuel market system, including the numbers of participants at each level, and the length of the market chain make it difficult in most cases to permit monopolistic practices by any single actor. Again, Shinyanga and Mbeya are less favourably structured than is Dar es Salaam, with larger and fewer traders operating in the two smaller towns. In Mbeya, the wholesalers are mostly small, but skewed by the existence of a few large wholesalers, while the retailers are mixed. In Shinyanga, on the other hand, both retailers and wholesalers tend to be large. Large numbers of participants limit potential profits of traders as a whole, even if some are more favourably positioned than their trading colleagues or producers.

Bonds between Participants

Despite this concentration of larger and fewer wholesalers in Shinyanga and Mbeya, there is sufficient differentiation and choice between sellers and buyers in the three cities, and the conduct of the system is reasonably efficient. Although personal bonds exist between buyers and sellers at all levels of the system, often for the purpose of facilitating trade and relations, they do not seem to be excessive in relation to the overall competitiveness of the system. There does not seem to be a competitive bottleneck because of them in either Mbeya or Dar es Salaam. This discounts the situation of nearly half of the producers in Mbeya who are hired by wholesalers to produce, and therefore are obligated to sell their production to that wholesaler. However, this falls more into the realm of integration than restriction.

What bonding does go on is mostly a result of lack of credit facilities, an acute problem in all three cities and in the informal sector as a whole. This sometimes requires credit to be given by a wholesaler to a retailer, or a wholesaler to a producer. This sort of relationship was not found to affect the fuel price adversely in any of the cities, though in Shinyanga two-thirds of wholesalers have a credit relationship with their sellers or buyers. This implies the sort of personal bond that has potentially constraining effects on the competitiveness of the market.

Ease of Entry

Ease of entry seems adequate for all three cities at the retail level. At the wholesale level, credit shortages are again a problem, though capital requirements are not excessive. Here, and also at the producer level, the greatest barriers to entry exist in Shinyanga and Mbeya. Harvesting licenses are rationed and this is probably a key reason for the concentration of wholesalers and the resultant constraints on competition in those cities. Though the reason behind the restricted licensing is to control the cutting of wood in the forest, the result is a concentration of legal cutters combined with fly-by-night illegal cutters. However, it is clearly easier to monitor a few large license holders than a large number of small licensees, and with the lack of resources available for monitoring this is a legitimate concern.

Length of Market Chain

Shinyanga and Dar es Salaam have the highest proportion of market chains that include only an end user, a trader and a producer. Clearly, the complaint heard that market chains in Africa are too long and filled with margin-grabbing and inessential participants is not true here. Mbeya is likely to include both a wholesaler and a retailer in its chain, and for firewood a three-node chain such as those in Dar es Salaam and Shinyanga is the norm. Despite this, the system integration is actually higher in Mbeya. There, wholesalers are generally in control of the production, although they tend to give up control of the retail function. Nowhere is the system so integrated that a large proportion of wholesalers owns the transport.

Conduct of Participants

Collusion

The conduct of participants in the market system is at times not ideal. Collusion among participants seems to be a potential problem among the charcoal wholesalers in Shinyanga, half of whom admit to setting prices together. The problem also exists within the large cooperative in Mbeya. Though many wholesalers in Mbeya are willing to discount their fuel below the price of their fellow sellers, no wholesalers in Shinyanga showed such an inclination. Combined with the credit relations Shinyanga wholesalers have with their customers and suppliers, this is a cause for concern. This situation may be possible in Shinyanga more than in the other cities because of the size of the market and the relatively small number of traders working in it. In Dar es Salaam, sheer size prevents much collusion. In fact, there is little evidence of collusion in Dar es Salaam or among participants at other levels of the system in any of the cities. Price matching is much more common among retailers than actual collusion, and it is often due to their marginal economic status that they do not strike out on their own by lowering prices. Price matching and collusion are particularly prevalent among Mbeya firewood sellers, but their marginality means that no individuals have market control. Predictably, they are also less likely than sellers in other cities to lower their prices.

System Performance

Economies of Scale

Producers In the realm of system performance, economies of scale exist at the producer level for charcoal both in the size of the kiln and the overall size of the producer. In both of these areas, Mbeya producers are found to be the most efficient. Despite this, Mbeya charcoal producers have the lowest margins and the lowest return to labour of producers in the three cities. This more than likely is tied to the fact that producers in Mbeya are funded and organized (in large groups) by the wholesalers there, and so have access to both labour and capital that eludes producers in the other areas. This is also the reason for their low returns, since the wholesalers absorb part of their potential margin. The implication here is that greater integration increases efficiency at the production level for charcoal, though it does not seem to have reduced the combined average wholesaler-producer margin. Instead, the wholesalers absorb the extra returns from this efficiency gain.

Sellers Charcoal sellers also realize scale economies. For wholesalers, larger wholesalers appeared to be more operationally efficient, while at the retail level medium-sized sellers performed better than their smaller or larger colleagues.

Transport Transport is subject to scale efficiencies, but overall transport efficiency is extremely low. This is due to the fact that trucks, the most commonly used form of transport, are generally sent out empty to pick up woodfuel at the site of harvesting. The organization of transport to allow haulage of goods in both directions would be a huge efficiency improvement for the system. The increased use of trains, a far more efficient form of transport, would also improve the system operational efficiency markedly. Transport is also a bottleneck to some extent in all three cities, but particularly so in Dar es Salaam where competing uses for even the poor quality vehicles used in woodfuel transport can find other uses. This, plus the high cost of repairs and the poor quality of the roads, contributes to the high transport costs, and pushes costs in Dar es Salaam above the other cities.

Spatial Arbitrage

These high transport costs and low transport efficiencies leave little room for spatial arbitrage. Though retail prices vary, wholesale prices of fuels are not distinct enough in different areas to encourage such arbitrage. If transport efficiencies are realized, and particularly if prices in towns with supply problems or inefficient or noncompetitive market systems go up significantly faster than those in towns with smoothly working systems, it is possible that there will be money to be made from spatial arbitrage in the future.

Margins

Despite the competitive advantages evidently experienced by charcoal wholesalers in Shinyanga and Mbeya, only the Mbeya traders seem to be profiting appreciably from it when their margins are appraised. Average Shinyanga wholesaler margins are the lowest of the three cities despite their seeming influence on the market. The potential excess profits are seemingly eaten up by high transport costs and higher operating costs, costs which the wholesalers seem unable to recover in the bag price of the fuel. Where they can recover costs and make profits is in retail sales due to the high prices for small volume sales, so those wholesalers who are also retailers (about two-thirds) make their profits there, negating the impact of low margins experienced at the wholesale level. Mbeya wholesalers are much less likely to involve themselves in large amounts of retail business, and they claim their profits at the wholesale level. At both wholesale and retail levels, however, the margins in the two cities are much higher than those in Dar es Salaam, which has a large market and seemingly few restrictions upon it. Returns to labour there are the highest of the three cities, despite the lower margins.

Annual Price Changes

When annual price changes are brought into play, real prices at the wholesale level in Mbeya decreased, while retail prices increased slightly, over the year studied. In Shinyanga, on the other hand, real prices at both the wholesale and retail level increased. These increases are undoubtedly partly due to increased transport costs, though increased profits are also likely. Real price increases in Dar es Salaam were moderate at all levels, though producers there had the largest real gains.

Historic Price Changes

Over a 15-year time span, real retail prices approximately doubled in both Mbeya and Shinyanga, while remaining constant in Dar es Salaam. This real price rise might indicate anything from price signals from rising fuel costs due to longer distances travelled by the fuel, to increasing system inefficiencies, to welfare losses caused by excess profits for wholesalers in a restrictively competitive system. In Dar es Salaam, prices have been mediated by competition with alternate fuels, but in Shinyanga and Mbeya this competition is less intense due to price and supply fluctuations for these fuels. In Mbeya and particularly in Shinyanga, increasing scarcity and the longer distances travelled by fuel are facts that surely raise the price of the fuels.

Producers

Producers are best off in Dar es Salaam, where they are independent and better compensated for their fuel. The restrictive licensing system in Mbeya and Shinyanga, combined with the fact that these licenses are likely to be held by wholesalers rather than producers (especially in Mbeya), add up to a lower average producer margin in those two cities. Returns to labour for producers in Shinyanga are the highest of the three cities, implying that perhaps the producers there may be better compensated than their margins imply. Some scepticism must be held for the hours per bag data, however, tempering any conclusions.

Firewood Margins

For firewood, margins are high in both Mbeya and Shinyanga at both the producer and trader level. Since many traders in Mbeya also collect the fuel themselves, the total margin is even higher, though the return on labour is very low. The lack of market size seems to hamper the competitiveness of the market in Shinyanga, in particular, leading to larger trader and producer margins.

Seasonal Price Changes

Seasonal changes in fuel prices were large in the year surveyed, but are moderate on a historical basis. Price increases are found at the producer level, in particular in Dar es Salaam, reflecting the decreased supply there as the farmers who make up the majority of the producers return to farming. In the other two cities, the existence of a higher number of full-time woodfuel harvesters mediated wet season price rises. Wholesalers of charcoal in Shinyanga and Mbeya are able to capture what seem to be margins over and above their increased seasonal costs, while the margins in Dar es Salaam actually decrease in the wet season because their extra costs are not covered. This is another indication of wholesaler influence in Shinyanga and Mbeya. Retailers have stable or decreasing margins in the wet season in all three cities. Other results have indicated that, for the three cities together, only medium-sized wholesalers and large and small retailers have increased margins in the wet season. Since most wholesalers in Shinyanga, and a good number in Mbeya fall into this 'medium' category, their extra wet season margins are confirmed. Sellers of firewood in those two cities also seem to capture excess profits in the wet season, particularly in Shinyanga, driven by fuel shortages.

Storage of Fuel

The high margins enjoyed by wholesalers in the wet season makes it unlikely that they would be interested in storing fuel from the dry season to the wet season. Traders in Dar es Salaam and retailers in all three towns, on the other hand, do not increase their profits substantially in the wet season and might therefore be interested in buying fuel in the dry season at lower prices and selling it at higher prices in the wet season. Producers, as well, might be able to profit from temporal arbitrage. However, very few do. The reasons are mostly related to the lack of capital or credit resources experienced by most retailers, small wholesalers, and producers. Most need to sell what they buy as soon as possible in order to have enough money to buy another stock of fuel to sell, and money to live with. In the long run, it turns out that temporal arbitrage is somewhat irrelevant. The historical seasonal rise in fuel prices is less than 15 per cent over the three cities, and in Dar es Salaam there is a historic drop in wet season prices. Therefore, the viability of seasonal storage, especially after risks and costs are incorporated, is minimal for sellers. This lack of viability does not extend to producers, but, except perhaps in the Dar es Salaam supply region, the price increases are also too small to justify the capital risk required.

Conclusions

This list of findings culled from the body of the paper leads us to a series of conclusions about the level of competitiveness and efficiency in each of the three cities, and, by extrapolation, the country as a whole. From there a set of policy recommendations can be advanced, and the experienced gained in this exercise used to suggest parallels in other developing countries, and areas for further research.

From the preceding discussion, a table for each fuel can be produced which lists the most important differences between cities to try to better distinguish the problems in each. From these tables we can distinguish the effects and results associated with the competitive and efficiency atmosphere in each town.

Dar es Salaam

Tables 8.1 and 8.2 show these differences. From the information organized in the tables, it is clear that Dar es Salaam, of the three cities, has the most competitive and efficient environment. No serious problems with the markets there have been identified, aside from efficiency losses that plague the system

144 *Woodfuel Markets in Developing Countries*

Table 8.1 Differences between charcoal markets in three cities

	Dar es Salaam	Mbeya	Shinyanga
Structure			
Concentration, length of market chain	Average length chain (3–4 nodes) No concentration of sellers Choice of sellers at all levels	Slightly longer chain (4 nodes) Concentration of larger wholesalers, esp. cooperative group controlling 20% of market Choice of sellers at all levels	Average length chain (3–4 nodes) Concentration of large wholesalers Concentration of larger retailers (often doubling as wholesalers) Less choice of sellers at retail and wholesale level
Vertical integration	All levels independent	Many producers employed by wholesalers Mostly professional producers	Some producers employed by wholesalers Some producers professional
Credit, personal bond between seller and buyer	Difficult to obtain No influence from buyer-seller credit relationships	Difficult to obtain Wholesalers employ producers	Wholesalers have credit relationship with two-thirds of buyers
Ease of entry	No extraordinary restrictions	Harvesting license required and rationed. Usually owned by wholesaler.	Harvesting license required and rationed. Sometimes owned by producer group.
Supply issues	—	Some scarcity pressures	Scarcity pressures
Conduct			
Collusion	No serious collusion	Cooperative group sets prices together	Half of wholesalers collude to set prices
Sales promotion	Price discounts offered at times	Price discounts offered at times	No discount
Performance			

Policy Implications and Conclusions   145

	Dar es Salaam	Mbeya	Shinyanga
Economies of scale	Wholesaler economies of scale	Producer economies of scale Limited wholesaler economies of scale	Limited wholesaler economies of scale
Allocative efficiency	Transport bottlenecks Inadequate use of trains Highest per km transport costs	Some transport bottlenecks	Some transport bottlenecks Inadequate use of trains Highest per bag transport costs
Margins	Highest producer margins Lowest retailer margins	Highest wholesaler margins Lowest producer margins	Lowest wholesaler margins Highest retailer margins—wholesalers are also retailers Highest for producers
Return to labour	Highest for wholesalers Highest for retailers	—	
Annual price movement	Moderate increases Largest gain by producers	Moderate wholesale decreases	Retail and wholesale increases
Historic real price trend	Stable	More than doubled in 15 years	Tripled in 15 years
Seasonal price movement	Wholesale margins decrease Producer price increases	Wholesale margins increase	Wholesale margins increase

Table 8.2 Differences between firewood markets in three cities

	Dar es Salaam	Mbeya	Shinyanga
Structure			
Market size	Large market	Medium market – many collect their own	Small market – many informal sector businesses
Concentration, length of market chain	Average length chain Larger retailers Choice of sellers at all levels	Slightly longer chain Concentration of larger wholesalers Concentration of small retailers Choice of sellers at all levels	Average length chain Concentration of large wholesalers Concentration of larger retailers Less choice of sellers at retail and wholesale level
Ease of entry	No extraordinary restrictions	Access to local fuel for harvest by small-scale sellers Harvesting license required and rationed for large wholesalers	Harvesting license required and rationed. Sometimes owned by producer group
Supply issues	—	Some scarcity pressures	Scarcity pressures
Conduct			
Collusion	No serious collusion	Collusion and price matching among retailers	No serious collusion
Sales promotion	Price discounts offered at times	Unwilling to discount	Price discounts offered at times
Performance			
Allocative efficiency	Transport bottlenecks Inadequate use of trains Highest per km transport costs	Some transport bottlenecks	Some transport bottlenecks Inadequate use of trains Highest per kg transport costs
Margins	—	High trader margins High producer margins	High trader margins High producer margins
Return to labour	—	—	Highest for producers and traders
Annual price movement	Moderate end user and producer price increases	Minor price increases	Largest end user and producer price increases
Seasonal price movement	Large producer price increases	—	Large end user price increase

as a whole. Dar es Salaam stands out as different for its independent producers and its well-functioning wholesale and retail markets. The size of the market in Dar es Salaam undoubtedly has a role in achieving this efficiency, since a large market gives room for many participants and makes it much more difficult for any one or any group of participants to exercise undue control. This competitiveness has been evident throughout this study.

Given the smooth working of the Dar es Salaam market, it can now be used as a base case against which the workings of the markets in the other two cities can be compared and contrasted. Cursorily, it appears that Mbeya follows Dar es Salaam in degree of efficiency and competitiveness, trailed by Shinyanga's market. Although first appearances implicated Mbeya's market as the least competitive of the three cities, further investigation has weighed the scale against Shinyanga.

Mbeya

Mbeya's charcoal market is medium-sized, as compared to the very large market that exists in Dar es Salaam. The market's distortions centre around the relationship between wholesalers and producers. The licensing system there limits licenses and they are usually owned and controlled by the wholesalers, and producers are employed by the wholesalers and are generally non-local professional producers. Primarily for this reason, wholesalers are able to receive margins higher than either of the other two cities. They also are able to collect increased margins during the price increases of the wet season. Although the 'cooperative' selling group sets their prices together, other wholesalers are very independent and collusion and market control, in a general sense, are not a problem. This is emphasized by the fact that the margins of the wholesalers actually decreased over the year of the study. Though the charcoal price in the city has doubled over the past 15 years, the rise can be attributed partly to a fast growing population and supply pressures. Though because of the licensing system in the city wholesalers are able to collect margins above what they would without the system, indications are that they have little if any actual control over the price of the fuel. This is because the numbers of wholesalers and retailers are sufficient, and the competition with other fuels intense enough to mitigate their influence. There is a benefit to their integration, in that the larger kilns that their control over and financing of the production process allows efficiencies of scale to be realized at that level, though at this point in time those efficiencies are not passed on to the consumer.

Mbeya is unique among the cities in its reliance on and access to firewood. The structure of its firewood market is very different from the other cities in its reliance on small-scale vendors aimed at the household market. The price of firewood is higher in Mbeya than in the other towns, and these retailers, who tend towards price matching, are at a level marginal enough and are numerous enough to lend some inefficiency to the market. It is possible that they actually keep prices up – since they can not lower their prices because their returns are already so low, other, larger, retailers that might otherwise lower their prices are able to keep them high.

Shinyanga

In Shinyanga, some of the same inefficiencies occur in the charcoal markets as in Mbeya, along with some additional problems. Shinyanga's market problems centre around the size of the charcoal market there. Because it is a small market, efficiencies dictate wholesalers who, while not large when compared to those in Dar es Salaam, are few enough in number to be able have some price control over the market. Half of the wholesalers admit to price-setting, and many have what seem to be exclusive credit-based relationships with buyers or producers. Again here, licenses are restricted, though there are more seasonal local producers here than in Mbeya and more licenses held by independent village and family groups. Margins for these wholesalers (most of whom are integrated forward as retailers) are higher both in the wet season and over the year which was studied. Prices in Shinyanga have tripled over a 15-year time period. There are definite supply pressures here, as population grows and overlapping supply areas increase demand, and fuel travels distances far greater than for the other cities, all of which contribute to the price increases and to the high cost of the charcoal in the town. It is not clear how much of the price rise is also attributable to market distortions caused by trader control. The lack of viable, affordable fuel substitutes also contributes to the high prices of charcoal.

The firewood market is more distinctly concentrated. With an even smaller market, much of which consists of informal sector merchants who are inflexible in their fuel choice, and with reducing supply areas, firewood producers and sellers are able to collect the largest margins of the three areas. Seasonal margin increases and annual increases were also large. Again, the size of the market for firewood in Shinyanga seems to be affecting its ability to run efficiently.

Policy Implications and Conclusions

Inefficiencies Found in all Towns

In all three areas transport is a large source of inefficiencies, and a bottleneck that may decrease the overall efficiency of the system as well. Illegal cutting is fairly common in all the areas, despite the license restrictions in Mbeya and Shinyanga, circumventing the licensing process. Though some of the illegal producers harvest trees from restricted areas, most of them cut in legal areas but avoid the quantity restrictions that the forest authorities have tried to enforce in an attempt to manage the forests.

No Evidence of Economic Naiveté

Though there are some wholesalers in the system who seem to be enjoying extra profits, the conditions feared in some of the literature, where wholesalers are taking advantage of economically naive small retailers, have not been found. Nor has any indication that decisions are being made by any of the players in the woodfuel system that may be considered anti-economic been encountered. In fact, the players involved have acted very recognizably like independent small business people anywhere in the world.

Economy of Affection

Personal bonds have been encountered as a player in the markets of all three cities. They seem to play a somewhat reduced role in Dar es Salaam as compared to the other cities. This may have to do with the more ingrained 'Western' commercial status of the capital city. Also, the 'economy of affection' described by Hyden (1980), although evident, may have been weakened by the separation of the participants from their local milieu into the anonymity of a large urban centre. Mbeya and Shinyanga are far smaller towns, with smaller immigrant populations. Those that have migrated to these towns are also more likely to have moved from a village in the surrounding areas. These factors make it more likely that the economy of affection is strong in Mbeya and Shinyanga, and this seems to have been borne out by the results of the study.

Patterns Observed

Four patterns have been established in this review. Some of the patterns described are ultimately the result of the low level of physical and economic infrastructure in the country, and would not hold in a country with higher levels of these essential components.

The first pattern observed is that the larger the city, the better its woodfuel delivery system operates in terms of efficiency and competitiveness. That vertical integration, in this case, seems to create inefficiencies in greater numbers than it relieves them is a second finding. A third is that independent producers, though somewhat less efficient than larger producers created in an integrated system, help to ensure the competitiveness of the system by working in their own interests. Fourth, supply pressure may be partly communicated through prices, but the inefficiency in the same markets seems to have a more significant effect on prices.

Supply Pressure and Prices

This fourth finding is important in that it links the study back into some of the issues brought out in the introductory chapter. Specifically, the externalities connected with woodfuel use and their effect on prices need to be re-examined. As was discussed in the introduction, woodfuels are cut at very low fees or for free. As a result, environmental externalities associated with the use of the woodfuel are not internalized into the price of the fuel faced by the end user. It is not evident from this study the exact relationship between the push upward on prices due to inefficiencies in the supply system and the push downward on prices by the absence of the reflection of externalities in prices. Although the supply pressures on the forest resource may have already communicated some upward pressure on prices in some areas, there is little indication that the full impact of environmental externalities is communicated in prices anywhere in Tanzania.

As was pointed out in the introduction, it is important to take into account the distortions imposed upon the supply market for woodfuels by both the distribution network and the externalities of supply. The two, as is becoming clear, cannot be completely separated. In general, as was brought out in the body of the study, harvesters of firewood and makers of charcoal are paying little in the way of cutting fees on public land in Tanzania. Although there are nominal stumpage fees assessed on the cutting of trees for different uses (timber, poles for construction, firewood, charcoal, etc.), none are high enough to fully reflect the environmental and social consequences of their use. They also are faulty in that they do not differentiate trees cut near cities from those cut far from areas of demand. The complexities of setting stumpage fees will be considered in more detail in the next section.

Contrast with Commercial Fuel Markets

The situation in woodfuel markets in Tanzania contrasts markedly with the situation prevailing in the markets for alternative fuels the so-called 'commercial' fuels such as electricity, kerosene, and LPG. These fuels are heavily regulated and controlled by the government through parastatals. Prices are controlled, and supply is often inadequate, leading to shortages and aftermarket sales at prices much higher than the official rates. In comparison, woodfuels are consistently available at fairly competitive prices. It is one of the few uncontrolled and unregulated markets in Tanzania.[1]

Comparison with other Countries

The Tanzanian situation can also be compared with the situation in Senegal, where entry barriers and influence allow wholesale prices and cutting quotas to be set by a cartel of *patrons*, primarily as a result of government interventions in the markets (Ribot 1988). It is not clear whether this has caused retail prices increases, but certainly trees are being cut in quantities and locations that they would not otherwise be. The Tanzanian experience as it has been described here also contrasts with the full integration and efficient transport that have kept prices low in Sudan. It contrasts as well with the situation in Ouagadougou, where traders and transporters have found that in recent years they have not been able to completely transfer producer price increases to end users due to price elasticities (Bertrand 1985). As a result they have taken a cut in their margins, a trend which may force traders and transporters there to be more efficient.

Policy Recommendations

This points away from direct government intervention in the woodfuel markets in Tanzania. Though efficiencies obviously exist in the market, specific problems need to be addressed individually, without a wholesale reworking of the system which, as was pointed out in the previous section, does the job of delivering fuel to urban dwellers much better than competing systems in the country.

The question at this point is what should be done to alleviate problems in the woodfuel markets in Tanzania as they stand. One problem with across the board policy recommendations is that fuelwood problems, both in supply and in market inconsistencies, tend to be local, so that national policies can be helpful in some towns and detrimental in others. However, there are some recommendations based on the patterns identified before that should have broad applicability. It is important to note that the workings of the market in a particular city should be understood before imposing changes, particularly because there may be some market types extant in the country that were not represented in this sample of three cities.

Efficiency Improvements

System-wide efficiency gains will help to improve the supply network in all of the cities. Previous sections have highlighted several areas in need of improvement. One place where efficiency gains certainly can be realized is in the organization of the woodfuel transport system to encourage back hauling. This would probably necessitate newer vehicles, since the lorries now being used are too unreliable to be part of such a system. Additional gains through vehicle fuel efficiency and a reduction in the 15 per cent to 20 per cent loss in charcoal due to the roughness of the handling and ride on the older trucks (as well as the rough roads) would also be realized.

Back hauling As discussed earlier, there is presently some back hauling in the system, but it is carried out at the whim of truck drivers, and generally involves the purchase of charcoal from roadside vendors. Since it occurs for the most part without the approval, in most cases, of the vehicle owner, if he is not driving. As such, the back hauling opportunities are limited because of their illicit nature. At the present time, many roadside sellers sell mostly to individuals in cars and trucks who are not primarily in the woodfuel business. Their value-added and margin makes it more profitable for full time wholesalers to go directly to the producers, despite added time and transport costs. The use of larger trucks on the main roads, and the reduced transport costs achieved in back hauling could alleviate the cost differences enough to make roadside pick up equally profitable or more profitable than kiln site pick ups.

For back hauling to have a significant effect on the system, organization at a high level and information sharing far beyond any presently attempted would be required. The system would have to be somewhat restructured, as woodfuel wholesalers and wholesalers of other products have to cooperate. The easiest solution would be for wholesale markets, which exist in a limited way in Dar es Salaam, to be strengthened and used as drop-off points for the back hauling transporters. These might also function as information clearinghouses, allowing transporters and woodfuel wholesalers to network and set deals. At present, woodfuel wholesalers travel with transporters to kiln sites to pick up fuel, as they are the ones who have made arrangements with producers, know the location of production sites, etc. In a back hauling scheme, this would be awkward if a transporter were taking a longer trip on the outward bound portion, and planned to stop at production sites after again nearing his city of origin. It would then require additional time for woodfuel wholesalers to either travel out of their way with the transporter as he delivered his goods, or found alternative transport for themselves to a point along the main road where the transporter could be met and led to the production sites. Delays by the transporter would incur additional time requirements on the wholesaler. These sorts of costs are probably, along with informational constraints and lack of cooperation by transport owners, what dissuade wholesalers from availing on the potential of back hauling now. The scheme of roadside pick up combined with wholesale centres in the cities is the most efficient organizationally to avail of back hauling efficiencies, but is enough of a structural shift that it is unlikely to occur easily.

In the Tanzanian context, these improvements would have a trade off in other ways. The transport of woodfuel provides a use for many vehicles that might otherwise, because of their age and condition, be unusable for goods transport. Newer trucks are presently not forced to travel over the extremely poor roads that characterize access to most kiln sites, and thus do not deteriorate as quickly. In a country that is under-supplied with commercial vehicles, these considerations must be weighed against the benefits of a more efficient transport regimen for woodfuel. The savings in imported petroleum engendered by back hauling woodfuel must be compared to the savings in imported vehicles and vehicle spares by using existing vehicles. With foreign exchange markets in the country newly opened, however, it is possible that new vehicles would be purchased by privately held foreign currency, as opposed to governmental reserves that are used to purchase petroleum.

However, it is quite possible that in a liberalized financial climate, there will be an increase in new commercial vehicle and vehicle spares that will facilitate the efficiency improvements detailed above. If the addition of new or larger trucks makes roadside pick up profitable, the older trucks now being used for woodfuel transport into the urban areas could be profitably used to provide transport between the production sites and the main roads while the newer and larger trucks remained on the main roads, reducing the wear and tear on them. This would also allow back hauling to be effectively practised, and the overall system efficiency would be greatly enhanced from all of the improvements combined. Both the facilitation of the purchase of trucks, and the improvement of roads will encourage this to happen, as entrepreneurs begin to see the profitability in the revised structure, and thus begin to change it.

Trains The increased use of trains, and encouragement of their legal use for woodfuel transport, as opposed to their current illicit use, would provide additional increases in transport efficiency. This improvement depends on the cost of adding to the rail infrastructure to allow for this additional cargo. Primarily, the adjustments needed in order to unload charcoal and firewood would require some planning and, perhaps, the purchase of additional cars. The most effective way to accomplish this would probably be to have depots at certain villages along the railway. At these depots, if cost effective, could be empty cars that would be loaded and ready for pick-up when the train arrived. This would require the addition of side rails in addition to extra cargo cars. Alternatively, the woodfuel could be piled near the tracks, and then loaded onto the train into an existing car when it arrived. The effectiveness and costs in time and labour of either or both of these possibilities would need to be better analysed, and the resulting charges for traders calculated. If they were not substantially lower overall, it may be difficult to convince traders to use the trains. Also required are changes in regulations allowing trains to stop to pick up fuel and to carry it, legally and with fees paid to the railway company rather than to employees working on the side.

Credit Another efficiency improvement needed, which is in fact a second but non-transport related infrastructural improvement, is the oft discussed improvements in credit availability. In this case it could help retailers and small wholesalers to remain independent, and all participants to enjoy short term price arbitraging, such as that which takes place over the seasonal changes. This is particularly useful in the case of producers, who might reduce supply fluctuations while improving their own earnings.

Local Control

Earlier in the chapter it was concluded that the production situation in Dar es Salaam is preferable to that in the other two cities. The producers there, independent but under the eye of village authorities as far as their land access, were better compensated, and the serious system constraints created by the licensing systems in Shinyanga and Mbeya were absent. Because of this, recommendations to create a situation in those cities' production areas emulating that in the Dar es Salaam area are in order.

Autonomous village regulation The first recommendation is for autonomous village regulation and revenue gathering for woodfuel harvesting. In all areas harvesters resent paying fees and taxes, primarily because they see no benefit from these moneys. If some of the fees were to stay within the communities that contain the forests being harvested, the incentive to enforce both fee-paying and management policies would be far greater. Even the illegal harvesting of reserves and other protected areas would be easier to monitor if responsible local authorities were protecting the fiscal and environmental interests of the village. In the Dar es Salaam area harvesting is done almost exclusively by village members on village land, facilitating local regulation, which would put in control of the land those who are most affected by its degradation.

Licensing producers Harvesting in Mbeya and Shinyanga is most frequently done by professional charcoal makers who are not tied to the land from which they are harvesting, In these areas, it is important to allow local authorities autonomy in granting harvesting licenses as well as in fee collection, allowing local residents to assert usufructuary rights, or auctioning them to outsiders if local producers are not sufficiently interested. Local residents should also be allowed to close a forest area to any harvesting at all, if they find that its degradation is deleterious to their way of life. This has been done in Ghana and Rwanda and it has helped to focus attention on and improve the efficiency and sustainability of the woodfuel system (Feinstein and van der Plas 1990). The closing of the supply areas did not, as expected, cause havoc due to shortages and huge price increases for users of woodfuels in the urban centres. Instead, the distribution system was forced to become more efficient.

From an efficiency standpoint, too, it could help to reduce the license bottleneck which allows control and large profits to be accrued to wholesalers, when they could be accruing to the local residents and urban consumers, while still returning fair compensation to the wholesalers. It is this bottleneck which has helped to cause much of the inefficiency in the Mbeya and Shinyanga markets. Apart from simply abandoning the licensing system altogether, the idea of local control seems to have the most promise as a system that will control access and preserve the habitat while avoiding competitive inefficiencies.

Recently, modifications to rural land tenure law have been considered (Hofstad 1997). If more secure tenure is extended to those areas which have seen reduced local control, it is possible that reduced impact might be seen on local lands.

Local money into local coffers Local control over revenue created by woodfuel harvesting would encourage collection of fees, support the personnel to do the collecting and the monitoring of cutters in the forest, as well as funnelling money into the local communities for seedling programs, village afforestation programs, and other local initiatives. Only if the money from the forest is reinvested in the forests will there be proper incentives and funding for sustainable maintenance of the woodlands.

Externalities and Stumpage Fees

Even in the face of local control, a problem arises in the wake of perceived supply problems in at least two of the areas studied in the inability of the price of the product to reflect the long-term environmental and social costs of its use.[2] Shinyanga has the most acute problem with deforestation, and it is also the least competitive of the three cities studied. Part of this lack of efficiency in the woodfuel markets there might be attributable to the fact that the local areas are deforested, and relatively long distances must be travelled to meet supply demands in the city.

The implications of the environmental costs of woodfuel harvesting, such as deforestation, go far beyond the provision of energy into areas such as:

- the preservation of topsoil and hence agricultural productivity;
- the protection of watersheds and the water supply; preservation of species diversity and faunal populations (and thus economic benefits from tourism);
- the urban-rural dichotomy whereby the rural dwellers bear the costs of the urban dwellers' consumption;
- loss of carbon dioxide fixing capability; and other issues.

It is not at all clear that the prices of the fuels are reflecting such hidden costs.

One possible policy implication is to apply higher taxes and fees to the fuel so it will more accurately reflect these externality costs, and to direct producers to cut in areas and use harvesting procedures that will do the least damage to the environment. The composition of taxes, unlike price controls or nationalization, preserves the competitive nature of the market while addressing the issue of externalities.

A problem with fees and taxes is that their collection is very difficult, especially in such a dispersed business as woodfuel harvesting. The fees that are collected now from woodfuel are thought to be at best about half of those owed (Silviconsult 1991). Increased collection of fees, and consistent application of fee schedules could help to regulate the system, and is a prerequisite to any fee revision action.

Setting stumpage fees Another problem with the revision of stumpage fees is how to determine a level which reflects the externalities of wood use. Data on environmental effects of deforestation are few, and difficult to generate in any case. According to the best available work on fuelwood stumpage (Openshaw and Feinstein 1989), three main methods are used to estimate stumpage values:

- market approaches (e.g. residual stumpage or auction sale/tender);
- surrogate market approaches (e.g. opportunity cost of collector's time); and
- cost-based approaches (e.g. fuel substitution, replacement cost).

Calculations in all cases can be done using either economic or financial costs and prices.

Market approaches Market approaches assume the market clearing price of fuel or standing stocks of wood reflect the full value of the resource. Residual stumpage calculates a stumpage value by taking the price of the woodfuel in the market and working backwards, accounting for all of the costs and profits of the intermediaries. Because transportation cost is a large part of the final cost of woodfuel, the stumpage value so calculated would reduce as distance from the market increases. Problems occur when the trees presently being cut are obtained free or for a very low price. The market price then does not reflect the externalities of woodfuel use. This points out the conflict between the market and social values of a resource. Since this analysis is making the assumption that there is value to a wood resource that is not captured by even an efficient market, a serious problem is posed by this particular method for determining the stumpage value.[3]

Auction sales assume full knowledge of opportunities and opportunity costs associated with a particular plot of standing stock, an assumption that is unlikely to be true in a developing country, and is certainly the case in Tanzania.

Governments selling standing wood stock tend to undervalue the wood because they do not realize the potential value in the wood (Openshaw and Feinstein 1989). Again, the market and social values of the wood may not be the same.

Surrogate market approaches Surrogate market approaches are useful when the product is not a marketed good. This is true for woodfuel in rural areas where people collect wood for fuel and there are no representative markets to guide the market price value of the woodfuel. In this case, the wood can be valued according to the non-market cost to the person who is collecting the wood. Since most of the cost of gathering wood is time, the suggestion is that the value of the wood can be estimated by valuing the labour opportunity cost of the gatherer's time. This method is appealing because in areas of scarcity, the value of the wood increases as the time taken to collect it increases. The rural minimum wage is usually used as a proxy for cost of labour. Unfortunately, this value is often not representative of the true opportunity cost of a rural person's time. Additionally, the labour opportunity cost for a rural worker in the growing season is far higher than that in the off season. In Tanzania, rural minimum wages lag far beyond inflation, and are unlikely to reflect the true value of rural labour.

Instead of using a minimum wage as a surrogate, the opportunities for productive labour in a particular place can be valued. However, this is extremely data intensive, and is difficult to do at any large scale.

This approach does no better at solving the market versus societal valuing of wood resources than the previous approaches. There may be opportunity costs for the resource that are not reflected in the labour opportunity costs. At best, the labour opportunity cost method for valuing woodfuel should be used to establish a lower bound for the value of the resource (Openshaw and Feinstein 1989).

Cost-based approaches Cost-based approaches to fixing stumpage values look at the value of supplying the product or a substitute product. The alternative fuel substitution method compares the value of a substitute fuel, usually kerosene, with that of woodfuel to determine whether the woodfuel price is competitive (Openshaw and Feinstein 1989). In Tanzanian cities, as Table 8.3 indicates, charcoal prices are lower than those of kerosene particularly when the economic price of kerosene is considered. Effective firewood prices are also lower than kerosene prices, although in Mbeya this only becomes apparent when the economic price for kerosene is used. If the price for woodfuels in Tanzania were derived using the substitute fuels method, stumpage prices

Table 8.3 Kerosene substitute price for woodfuel

	Market price (Tsh/unit)	Unit	Energy content (MJ/unit)	Price/energy unit (Tsh / eff. MJ)	Typical efficiency (%)	Effective price (Tsh/ effective MJ)
Dar es Salaam						
Firewood	6.1	kg	15.5	0.39	10	3.94
Charcoal	20.8	kg	29.0	0.72	20	3.59
Charcoal (improved stove)	20.8	kg	29.0	0.72	30	2.39
Kerosene	55	l	35.0	1.57	30	5.24
Kerosene (econ. price)	95.9	l	35.0	2.74	30	9.13
Mbeya						
Firewood	10.9	kg	15.5	0.70	10	7.03
Charcoal	21.7	kg	29.0	0.75	20	3.74
Charcoal (improved stove)	21.7	kg	29.0	0.75	30	2.49
Kerosene	55	l	35.0	1.57	30	5.24
Kerosene (econ. price)	114.4	l	35.0	3.27	30	10.90
Shinyanga						
Firewood	8.8	kg	15.5	0.57	10	5.68
Charcoal	25.2	kg	29.0	0.87	20	4.34
Charcoal (improved stove)	25.2	kg	29.0	0.87	30	2.90
Kerosene	55	l	35.0	1.57	30	5.24
Kerosene (econ. price)	109.7	l	35.0	3.13	30	10.44

Source: Hosier and Kipondya 1993. The economic price for kerosene takes into account the duties and cross-subsidies on that fuel.

could be increased substantially to match the substitute price of kerosene.

One problem with this method is its sensitivity to exchange prices, stove efficiencies, and other numbers used in calculation. Additionally, kerosene is not a perfect substitute for woodfuels, and woodfuels must be given some value for the properties that make them a preferred fuel for many users.

A method that has been widely suggested is to base the stumpage fees on the replacement value, *via* plantations, of the tree that is cut. In Tanzania, the replacement cost of wood was estimated at $2.00 per m^3 in 1986 (Ahlback 1989) and $5.00 per m^3 in 1990 (Hosier and Kipondya 1993). Using the $5.00 per m^3 figure, Hosier and Kipondya (1993) calculated the economic cost of woodfuel. These are presented in Table 8.4. They are significantly higher than financial prices in the case of charcoal, even given the fact that they do not include the cost of environmental externalities (Hosier and Kipondya 1993).

There are problems with this approach as well. Since in some areas the regrowth rate of cut trees, if left alone, is significant, this might in some cases impose fees that are in excess of the true replacement value of the tree. In addition, the causes of deforestation in a locality can make a difference in the economically correct local stumpage fee. If charcoal making is an offshoot of agricultural clearing in an area, for example, the value of the tree to the farmer is negative. In this case the tree is a nuisance rather than a resource, and the replacement cost of such a tree is irrelevant if agriculture is the best use for the land. As well, controlling charcoal production will not significantly alter the forest usage patterns in the locality.

Table 8.4 Economic costs of woodfuel

	Economic cost (Tsh/effective MJ)	Financial price (Tsh/effective MJ)
Dar es Salaam		
Firewood	5.27	3.94
Charcoal	5.64	3.59
Charcoal (improved stove)	3.76	2.39
Mbeya		
Firewood	5.27	7.03
Charcoal	5.64	3.74
Charcoal (improved stove)	3.76	2.49
Shinyanga		
Firewood	5.27	5.68
Charcoal	5.64	4.34
Charcoal (improved stove)	3.76	2.90

Source: Hosier and Kipondya 1993.

Stumpage fees imposed on all cut trees, though, not just on those cut specifically for sale and not on the finished product, would require farmers to sell the trees they clear. This would clearly be infeasible in areas where there is no market for woodfuel, but such a stumpage fee could be formulated to reflect environmental and other damage done by the land clearing and agricultural use (Teplitz-Sembitzky and Schramm 1989). Clearly, the enforcement of, and setting of such fees would be cumbersome and difficult. However, the participation of local authorities would greatly ease such problems.

A per bag stumpage fee, though easier to enforce, does not necessarily encourage efficient production by producers. If a per bag tax is assessed, the incentive to use less cut wood per bag of charcoal is dependent on the physical and time benefits of having to cut and stack fewer trees, but not on any specific monetary incentive. In forested areas, during seasons of labour surplus (when most charcoaling is done), this incentive is minimal. However, monetary incentive is enough to encourage most producers to adopt or at least investigate efficient charcoaling technology. If stumpage fees are assessed per bag, the inefficient producer is charged less per kilo or stacked m^3 for the wood he uses than the efficient producer.

Stumpage fees must also reflect the distance of the fuel from the urban market. Because of transport costs, the value of standing wood lessens the further it is from the market. This was demonstrated in Figure 7.3. As well, additional fees representing a charge for the environmental costs of the particular harvesting site would help to account unbiasedly for those externalities (Templitz-Sembitzky and Schramm 1989).

Effects of stumpage fee changes Clearly, there is no definitive method to set stumpage fees. All of the methods presented here have flaws, and none effectively encompass the environmental externalities posed by woodfuel use. Although the idea of higher stumpage fees is sound, their implementation will be difficult for the time being.

Any change in the fee structure applied to woodfuels will affect the price of the fuel to the end user, and consequently affect the fuel choice of consumers. If the price of the fuel accurately reflects its cost to society, the government should be indifferent as to whether consumers choose the fuel. Although in theory, a correctly determined stumpage fee would only reduce the excess profits of suppliers, it is possible that much of the increase in fees would be passed on to consumers in higher retail prices. In Ouagadougou retail prices increases did not fully reflect increased costs to the traders because consumers began to switch to alternative fuels (Bertrand 1985). The sellers had to become more efficient or reduce their profit margins. In a system that is already extremely efficient and in which sellers have low margins to begin with, or in places where energy substitutes are not available, retail prices would increase at a pace with stumpage fee increases. In most countries, however, the indication is that retail price increases may reflect only part of stumpage fee increases.

In any case, programs such as improved charcoal stoves and improved kilns will serve to help the consumer reduce fuel expenditures, since the per tree replacement cost should be the basis for a stumpage fee. An obvious offshoot of a raised stumpage fee is that it should encourage afforestation, either as agroforestry or in plantations. Again, this is particularly true if moneys from woodfuel extraction are recycled through the local community. This policy has been enacted in Malawi, with some success (World Bank 1987a).

As Hofstad argues (1997), it is possible that, no matter how 'accurate' the stumpage fees are, they may be more expensive to enforce than the benefits from their enforcement will accrue. This is not a certainty, however, especially if the nonmonetary benefits of aspects such as reduced deforestation are included in the calculation of benefits.

Creation of Catchment Areas

A final issue is the protection of particularly endangered areas by the creation of more catchment areas and forest reserves. Though encroachment exists, the difference between areas that are protected and those that are not is often startling, and it is clearly an effective way to preserve forests, especially if the local residents are educated as to the benefits that it will bring them. This policy has been initiated in Rwanda, and found to be very effective (Leach and Mearns 1988, World Bank 1987a).

What Does it Mean?

Tanzania, as is the case with some African countries, has a functioning and, in fact, flourishing woodfuel supply system that serves a large proportion of the urban population, and provides income for a not insignificant sector of the rural and urban population. Though there is no woodfuel crisis per se in Tanzania, the health of the woodfuel system and the environment there is dependent on a good understanding and management of the system.

The first management tool required is information, and the project on which this article is based is a large step towards its provision. With this information, improvements can be made which will increase the efficiency and sustainability of what is, at its essence, a workable and working system.

The woodfuel market system in Tanzania is clearly a dynamic part of the informal economy, with many entrants and large numbers of participants overall. Most of the participants look to the woodfuel business as their primary source of income, and as such will be greatly affected by any fluctuations in the market due to policy changes. However, they will benefit overall by the improvements suggested here.

Though some of the suggestions detailed above may be politically unpopular, they can be made less so by judicious and gradual implementation and information dissemination. If the policy intentions of the government are made clear, participants in the woodfuel system and consumers alike will have an idea of what to expect in the future, and can plan accordingly. Setting of new stumpage fee structures, local participation in the fee setting and collecting and forest management, and the improvement of inefficient aspects of the system as it stands, will go a long way toward ensuring the longevity of woodfuel as a resource and as a major energy source for the country. It will also ensure the continuance of the delivery system of these necessary and popular fuels.

Broader Contribution and Future Work

The type of data gathered, the methodology used in gathering the data, and the manner in which the data was analysed in this study are all departures from previous studies of woodfuel in developing countries. As discussed in the first chapter, few studies have been done which examine the workings of woodfuel markets, and even fewer in any detail. This economic examination done on the Tanzanian woodfuel markets, using the structure, conduct performance paradigm, has provided a more in depth, socioeconomic look at these markets than previously done, with the exception of Ribot's work in Senegal, which took a political economic view of the woodfuel markets there (Ribot 1988).

The SCP paradigm, used extensively in industrial economics, and adopted for the examination of subsistence agricultural systems in Africa, was a reasonable basis for a study such as this one. Despite this, there were some techniques used in the agricultural studies from which the version of the paradigm used here was adapted (Jones 1972, Southworth et al. 1979, and others) which did not adapt well to the peculiarities of a woodfuel market. This was particularly true for spatial measures of arbitrage and integration. Woodfuel is rarely a candidate for export since it is relatively massive and expensive to transport. Therefore, there is little scope for countrywide spatial arbitrage, as the transport costs would severely outweigh the value of the fuel. On a more local basis, however, there are overlapping supply areas, as discussed earlier, which do provide arbitrage opportunities. If neighbouring towns were assessed, spatial arbitrage issues would increase in importance.

Another constraint on analysis of woodfuel markets was the lack of widespread historical price data. Because of its informal status in most countries, which has protected it from most regulation, few records have been kept on woodfuel prices, in particularly at the wholesale level, where records are essentially nonexistent in most, if not all, developing countries. This contrasts with the often poor, but at least existent, records for the agricultural sector, which has a long history of governmental intervention and interest. Though these lacks changed the type of analysis attempted, they did not prevent the paradigm from being relevant to the problem at hand.

The methodology used to gather data worked well, and, as explained in chapter 2, gave a more unbiased sample of woodfuel market participants than in previous surveys. For detailed examinations of distances travelled by the fuel it was also particularly well suited, as compared to other methodologies. The greatest improvement that can be made on the methodology (in addition to some improvements in the survey instruments themselves) would be to complement it with estimates of numbers of participants at each level, since it does not make those estimates well. A 'road block' survey of the volume of fuel entering a city would also complement household survey data on total consumption, and cross check volume estimates for supply areas made using the present survey methodology.

The survey instruments were designed to gather information about both attitudes and business practices, and both economic and non-economic data. Through this, detailed economic data relating to the business practices and structure of the market could be combined with the human aspects of doing business and buying energy products in a developing country. The result was a much more complete picture of the actual situation in the market there than with methods used previously. In this, the methodology worked well.

The survey instruments had several flaws that would, if rectified, add to the quality of the data and increase the analysis possibilities. These improvements are made more evident after the analysis of the data has identified areas in the system that are particularly important.

Firstly, more information about transportation arrangements, costs and contracts is needed. This would allow more precise discussion about the role of transport in the system, and give more information about constraints in the system at that level.

A second improvement in the surveys would be to acquire more detailed information on personal bonds and noncontractual arrangements between buyers and sellers of fuel, particularly between sellers and producers. Greater confidence would then be possible about such relationships, their duration, and their flexibility, leading to a greater understanding of their influence on the market.

Additional data about costs, both monetary and nonmonetary, facing sellers would also be helpful. Through this additional data true profits could with confidence be estimated, as opposed to the tentative margins estimated in this study. From this these middlemen could say more, in a more precise way, about the existence of excessive returns.

Finally, the surveys would benefit from more precise measuring of firewood quantities. This is a common problem in woodfuel surveys, and one that was not surmounted in the case of firewood in this one. Firewood does not grow nor is sold in standardized quantities, and it is extremely difficult to generalize even within the output of the same producer or seller. Although some weighing was done, the enumerators were not able to quantify adequately firewood volumes, making all estimates based on these prone to large errors. The problem was not encountered for charcoal, which is easier to weigh, and easier to generalize for an individual participant.

The study's emphasis on an economic and social analysis of woodfuel markets was unique. In most African countries, as well as many developing countries around the world, woodfuel is still a dominant fuel in the urban areas. In general, as is the case in Tanzania, woodfuel markets are little understood, and relegated to the informal sector of the economy, while 'commercial' fuels are highly tracked, regulated, and otherwise studied and controlled. This study has taken a first try at understanding the fuelwood markets and their economic and social underpinnings, along with the realities of their operation and sustainability.

The methodology followed here could reasonably be applied to other countries and other urban areas. Since the structure of woodfuel markets vary considerably from city to city, as we have seen, as well as from country to country, some idea of the basic structure of the woodfuel system in a particular place would be a useful starting point. A set of survey instruments following the examples in the Appendix can then be developed and administered. The use of the backwards-linkage methodology in administering the survey will provide the benefits described earlier, and is adaptable to any market structure.

As an analytical tool, the SCP paradigm, and the types of tests and analyses followed in this study, is also adaptable to a particular country- or city-specific set of circumstances. Through them, the efficiency and competitive nature of the woodfuel market in more cities and more countries can be analysed. With such additional studies, there will develop a body of work. It will then be particularly interesting to be able to analyse a number of studies of the type accomplished here to determine generalities about woodfuel markets. There may be similarities among woodfuel markets in countries or continents; between cities of the same size; between markets with similar supply access; or among all of the preceding. Comparisons may also be attempted to link woodfuel market studies with the body of work that has examined influences on woodfuel prices in the developing countries. There is also room for comparisons with studies of other developing country market systems, particularly those of staple food crops. Similarities may then be linked to market types rather than to the particular commodity being sold. In this way will contributions be made not only to the specific literature on woodfuel, but also to the more general literature on markets and their functioning.

Notes

1 As discussed earlier, a similar situation has been identified, especially of late, in the informal agricultural marketing sector. See e.g. Lynch 1994.
2 For a detailed look at this issue, see Hosier (1993).
3 For a detailed discussion of the question of whether woodfuel *should* be valued differently by market and society, refer to Openshaw and Feinstein (1989).

Bibliography

Adelman, M.A. (1955), 'Concept and statistical measurement of vertical integration', in G.J. Stigler (ed.), *Business Concentration and Price Policy*, Princeton University Press: Princeton, NJ.

Ahlback, A.J. (1988), *Forestry for Development in Tanzania*, International Rural Development Centre, Swedish University of Agricultural Sciences, Working Paper No. 71, Swedish University of Agricultural Sciences: Uppsala.

Alam, M., Dunkerley, J., Gopi, K.N., Ramsey, W. and Davis, E. (1984), *Fuelwood in Urban Markets: A Case Study of Hyderabad*, Concept Publishing Company: New Delhi.

Alam, M., Reddy, A. and Dunkerley, J. (1985), 'Fuelwood use in the cities of the developing world: Two case studies from India', *Natural Resources Forum*, Vol. 9, No. 3.

Alexander, J. and Alexander, P. (1987), 'Striking a bargain in Javanese markets', *Man*, Vol. 22, pp. 42–68.

Allen, J.C. (1985), 'Soil response to forest clearing in the United States and the tropics: geological and biological factors', *Biotropica*, Vol. 17, pp. 15–27.

Anderson, D. (1986), 'Declining tree stocks in African countries', *World Development*, Vol. 14, pp. 853–64.

Bain, J. S. (1959), *Industrial Organization*, John Wiley and Sons: New York.

Barrett, C. B. (1997), 'Food marketing liberalization and trader entry: evidence from Madagascar', *World Development*, Vol. 25, pp. 763–77.

Belshaw, C. S. (1965), *Traditional Exchange and Modern Markets*, Prentice-Hall, Inc.: New Jersey.

Berg, E. (1980), 'Reforming grain market systems in West Africa: a case study of Mali', in ICRISAT, Proceedings of the International Workshop on Socioeconomic Constraints to Development of Semi-Arid Tropical Agriculture, 19–23 February 1979, Pantancheru, India.

Bertrand, A. (1985), 'Market networks for forest fuels to supply urban centers in the Sahel', *Rural Africana*, Vol. 23–24, Fall 1985–Winter 1986, pp. 33–47.

Bevan, D., Bigsten, A., Collier, P. and Gunning, J.W. (1987), *East African Lessons on Economic Liberalization*, Thames Essay No. 48, Trade Policy Research Centre. Gower: Aldershot.

Boyer, N.A., and Davis, C.G. (1990), 'Exploitation and inefficiency in Cameroon food marketing systems – myth or reality?: Some evidence from the West Province', *The Review of Black Political Economy*, Vol. 18, No. 4.

Bressler, R.G., and King, R.A. (1970), *Markets, Prices, and Interregional Trade*, John Wiley and Sons: New York.

Cecelski, E. (1985), *The Rural Energy Crisis, Women's Work and Basic Needs: Perspectives and Approaches to Action*, ILO Rural Employment Policy Research Program, Technical Cooperation Report: Geneva.

CILSS/Club du Sahel (Working Group on Market, Price Policy, and Storage) (1977), *Market, Price Policy and Storage of Food Grains in the Sahel: A Survey*, Center for Research on Economic Development, University of Michigan: Ann Arbor.

Cline-Cole, R.A. (1984), 'Towards an understanding of man-firewood relations in Freetown (Sierra Leone)', *Geoforum*, Vol. 15, No. 4, pp. 583–94.

Cline-Cole, R.A. (1987), 'The socio-ecology of firewood and charcoal on the Freetown peninsula', *Africa*, Vol. 57, No. 4, pp. 457–97.

Davis, W. (1973), *Social Relations in a Philippine Market*, University of California Press: Berkeley.

Delgado, C.L. (1986), 'A variance components approach to food grain market integration in Northern Nigeria', *American Journal of Agricultural Economics*, Vol. 68, No. 4, pp. 970–9.

Dunkerley, J. and Gopi, K.N. (1985), 'Fuelwood Markets in Hyderabad', in *Agricultural Markets in the semi-arid tropics: Proceedings of the International Workshop*, 24–28 October 1983, International Crops Research Institute for the Semi-Arid Tropics: Patancheru, India.

Eicher, C.K. and Baker, D.C. (1982), 'Research on agricultural development in Sub-Saharan Africa: a critical survey', Working Paper, Michigan State University: Ypsilanti.

Ellis, G. (1981), 'The backward bending supply curve of labor in Africa: models, evidence, and interpretation – and why it makes a difference', *The Journal of Developing Areas*, Vol. 15, January, pp. 251-74.

Ellis, G. (1988), 'In search of a development paradigm: two tales of a city', *The Journal of Modern African Studies*, Vol. 26, No. 4, pp. 677–83.

Epstein, T.S. (1982), *Urban Food Market and Third World Rural Development*, Croom Helm: London.

ESMAP (1990), 'Examples of recent household energy survey work', internal memorandum, World Bank: Washington, D.C.

Feinstein, C. and van der Plas, R. (1991), *Improving Charcoal Production in the Traditional Rural Sector*, World Bank Industry and Energy Working Paper, Energy Series Paper No. 38, July, World Bank: Washington, D.C.

Foley, G. (1988), 'Discussion paper on demand management', in Proceedings of the ESMAP Eastern and Southern Africa Household Energy Planning Seminar, World Bank: Washington, D.C., pp. 25-50.

French, D. (1984), 'The economics of bioenergy in developing countries', in *Bioenergy '84*, Elsevier Applied Science Publishers: London.

Geertz, C. (1978), 'The bazaar economy: information and search in peasant market', *American Economic Review*, Vol. 68, No. 2, pp. 28–32.

Goldin, L. R. (1986), 'Organizing the World Through the Market: A Symbolic Analysis of Markets and Exchange in the Western Highlands of Guatemala', unpublished doctoral dissertation, State University of New York at Albany.

Harriss, B. (1979), 'There is method in my madness: or is it vice versa? Measuring agricultural market performance', *Food Research Institute Studies*, Vol. 17, No. 2, pp. 197–218.

Harriss, B. (1981), *Transitional Trade and Rural Development*, Vikas Publishing House: New Delhi.

Harriss, B. (1982), *Agricultural market in the semi-arid tropics of West Africa*, ICRISAT, Development Studies Occasional Paper No. 17, March: Hyderabad.

Harriss, B. (1984), *State and Market: State Intervention in Agricultural Exchange in a Dry Region of Tamil Nadu, South India*, Concept Publishing Co.: New Delhi.

Hays, H.M. (1975), 'The Market and Storage of Food Grain in Northern Nigeria', IAR, Samaru Miscellaneous Paper No. 50, Ahmandu Bello University: Zaria, Nigeria.

Hays, H.M. and McCoy, J.H. (1978), 'Food Grain Market in Northern Nigeria: Spatial and Temporal Performance', *The Journal of Development Studies*, Vol. 14, No. 2, pp. 182–92.

Hofstad, O. (1997), 'Woodland deforestation by charcoal supply to Dar es Salaam', *Journal of Environmental Economics and Management*, Vol. 33, pp. 17–32.

Hollier, G. P. (1986), 'The marketing of gari in North-west Province, Cameroon', *Geografiska Annaler*, Vol. 68B, No. 2, pp. 59–68.

Hosier, R. (1992), 'Energy use in Tanzania's urban informal sector: efficiency and employment potential in three cities', Working Paper, Stockholm Environment Institute: Stockholm, Sweden.

Hosier, R. (1993), 'Charcoal production and environmental degradation: environmental history, selective harvesting, and post-harvest management', *Energy Policy*, Vol. 21, No. 5, pp. 491–509.

Hosier, R., Boberg, J., Mwandosya, M. and Luhanga, M. (1990), 'Energy planning and wood balances: sustainable energy futures for Tanzania', *Natural Resources Forum*, May, pp. 143–54.

Hosier, R. and Kipondya, W. (1993), 'Urban household energy use in Tanzania: prices, substitutes and poverty', *Energy Policy*, Vol. 21, No. 5, pp. 454–73.

Hyden, G. (1980), *Beyond Ujamaa in Tanzania*, University of California Press: Berkeley.

Jayne, T.S. and Jones, S. (1997), 'Food marketing and pricing policy in Eastern and Southern Africa: a survey', *World Development*, Vol. 25, No. 9, pp. 1505–27.

Jolly, C.M. (1989), 'Pricing and selling decisions in a labor surplus economy', *Journal of Asian and African Studies*, Vol 24, No. 3–4, pp. 188–98.

Jones, W.O. (1972), *Market Staple Food Crops in Tropical Africa*, Cornell University Press: Ithaca.

Jones, W.O. (1974), 'Regional analysis and agricultural market research in tropical Africa: concepts and experience', *Food Research Institute Studies in Agricultural Economics, Trade and Development*, Vol. 13, No. 1, pp. 3–28.

Jones, W.O. (1987), 'Food-crop market boards in tropical Africa', *The Journal of Modern African Studies*, Vol. 25, No. 3, pp. 375–402.

Kaale, B.K. (1983), *Tanzania Five Year National Afforestation Plan*, Ministry of Natural Resources and Tourism: Dar es Salaam.

Karch, E., Boutette, M. and Christopherson, K. (1987), 'The Casamance Kiln', Energy Development International for USAID: Washington, D.C.

Katerere, Y. (1984), 'Issues in household energy strategy formulation: the Zimbabwe experience', in *Proceedings of the ESMAP Eastern and Southern Africa Household Energy Planning Seminar*, World Bank: Washington, D.C.

Katzin, M. F. (1960), 'The business of higglering in Jamaica', *Social and Economic Studies*, Vol. 9, No. 3, pp. 297–331.

Klitgaard, R. (1991), *Adjusting to Reality: Beyond 'State Versus Market' in Economic Development*, ICS Press: San Francisco.

Leach, G. (1987), *Household Energy in South Asia*, Elsevier: New York.

Leach, G. (1988), 'Interfuel substitution', in *Proceedings of the ESMAP Eastern and Southern Africa Household Energy Planning Seminar*, World Bank: Washington, D.C., pp. 25–50.

Leach, G. (1992), 'The energy transition', *Energy Policy*, Vol. 20, No. 2, pp. 116–23.

Leach, G. and Mearns, R. (1988), *Beyond the Woodfuel Crisis: People Land and Trees in Africa*, Earthscan: London.

Lele, U.J. (1971), *Food Grain Market in India: Private Performance and Public Policy*, Cornell University Press: Ithaca.

Lewis, L.A. and Berry, L. (1988), *African Environments and Resources*, Unwin Hyman: Boston.

Luhanga, M.L. and Kjellstrom, B. (1988), *Potential for use of biomass derived fuels in Tanzania* (Draft), Beijer Institute: Stockholm, Sweden.

Lynch, K. (1994), 'Urban fruit and vegetable supply in Dar es Salaam', *The Geographical Journal*, Vol. 160, No. 3, pp. 307–18.

Macauley, M., Naimuddin, M., Agarwal, P.C. and Dunkerley, J. (1989), 'Fuelwood use in urban areas: a case study of Raipur, India', *The Energy Journal*, Vol. 10, No. 3, pp. 157–80.

Maguay, M. and Makbel, M. (1990), 'Energy Survey in Private Households in Tanzania', Central Statistical Bureau, for the Tanzania Urban Energy Project: Dar es Salaam, Mimeo.

Mazambani, D. (1984), 'Commodification of woodfuel in Zimbabwe's urban areas: a case study of Harare and Chitungwiza, 1980–1984', Zimbabwe Energy Accounting Project Working Paper 9, August: Harare.

McGranahan, G. (1986), 'Searching for the Biofuel Energy Crisis in Rural Java', unpublished doctoral dissertation, University of Wisconsin-Madison.

McGrath, W.B. (1989), 'The challenge of the commons: the allocation of non-exclusive resources', Environment Department Working Paper No. 14, World Bank: Washington, D.C.

Mercer, D.E., and Soussan, J. (1992), 'Fuelwood problems and solutions', in Sharma, N.P. (ed.), *Managing the World's Forests*, Kendall-Hunt Publishing Co. for the World Bank: Dubuque, Iowa.

Mintz, S. (1961), 'Prat k: Haitian personal economic relationships', *Proceedings of the 1961 Annual Spring Meeting of the American Ethnological Society*, pp. 54–63.

Miracle, M.P. (1968), 'Market structure in commodity trade and capital accumulation in West Africa', in Moyer, R. and Hollander, S.C. (eds), *Markets and Market in Developing Economies*, Richard D. Irwin: Homewood, Illinois.

Munslow, B., Katerere, Y. Firf, A. and O'Keefe, P. (1988), *The Fuelwood Trap*, Earthscan Publications: London.

Norvell, D.G. and Thompson, M. K. (1968), 'Higglering in Jamaica, and the mystique of pure competition', *Social and Economic Studies*, Vol. 17, No. 4, pp. 407–16.

Nyanteng, V.K. and van Apeldoorn, G.J. (1971), 'The farmer and the market of foodstuffs', Institute of Statistical, Social and Economic Research Technical Publication No. 19, University of Ghana: Legon.

O'Keefe, P. and Munslow, B. (1988), 'Resolving the irresolvable: the fuelwood problem in Africa', in *Proceedings of the ESMAP Eastern and Southern Africa Household Energy Planning Seminar*, World Bank: Washington, D.C., pp. 25–50.

Openshaw, K. (1984), 'Tanzania', in O'Keefe, P. and Munslow, B. (eds), *Energy Development in Southern Africa*, Beijer Institute: Stockholm, Sweden.

Openshaw, K. (1989a), 'Zambia woodfuel market survey – rural road-side traders', draft working paper, World Bank: Washington, D.C.

Openshaw, K. (1989b), 'Zambia woodfuel market survey – urban market traders', draft working paper, World Bank: Washington, D.C.

Openshaw, K. (1989c), 'Zambia woodfuel market survey – woodfuel transportation and distribution study' draft working paper, World Bank: Washington, D.C.

Openshaw, K. and Feinstein, C. (1989), *Fuelwood Stumpage: Considerations for Developing Country Energy Planning*, Energy Series Paper No. 16, World Bank: Washington, D.C.

Perry, M.K. (1989), 'Vertical integration: determinants and effects', in Schmalensee, R. and Willig, R.D. (eds), *Handbook of Industrial Organization*, Elsevier Science Publishers: Amsterdam.

Plattner, S. (1983), 'Economic custom in a competitive marketplace', *American Anthropologist*, Vol. 85, No. 4, pp. 848–58.

Raju V.T., and Bhatt, B.D. (1985), 'Efficiency in pricing and operations of markets for semi-arid tropical crops in India – a case of groundnut in Gujarat', in *Agricultural Markets in the semi-arid tropics: Proceedings of the International*

Workshop, 24-28 October 1983, International Crops Research Institute for the Semi-Arid Tropics: Patancheru, India.

Ribot, J. (1988), 'Market structure and environmental policy in Senegal's charcoal industry', *Environment Africaine*.

Ruttan, V. W. (1969), 'Agricultural product and factor markets in Southeast Asia', Anschel, in K. R. et al (eds), *Agricultural Cooperatives and Markets in Developing Countries*, Frederick A. Praeger: New York.

Saul, M. (1987), 'The organization of a West African grain market', *American Anthropologist*, Vol. 89, No. 1, pp. 75–95.

Schmidt, G. (1982), *The Efficiency of Market Systems for Agricultural Products in Pakistan's Punjab*, Socio-Economic Studies on Rural Development, Vol. 14: Saarbrücken.

Semboja, J. and Rugumisa, S.M.H. (1988), 'Price control in the management of an economic crisis: the National Price Commission in Tanzania', *African Studies Review*, Vol. 31, No. 1, pp. 47–66.

Sherman, J. R. (1985), 'Food-grain market in Burkina Faso', in *Agricultural Markets in the semi-arid tropics: Proceedings of the International Workshop*, 24–28 October 1983, International Crops Research Institute for the Semi-Arid Tropics: Patancheru, India.

Silviconsult (1991), 'Forest revenue collection in Tanzania', mimeo, August.

Sosnick, S.H. (1968), 'Toward a concrete concept of effective competition', *American Journal of Agricultural Economics*, Vol. 50, pp. 827–53.

Soussan, J. (1988), *Primary Resources and Energy in the Third World*, Routledge: London.

Soussan, J., O'Keefe P. and Munslow, B. (1990), 'Urban fuelwood: challenges and dilemmas', *Energy Policy*, Vol. 18, No. 7, July/August.

Southworth, V.R., Jones W.O. and Pearson, S.R. (1979), 'Food crop market in Atebubu District, Ghana', *Food Research Institute Studies*, Vol. 17, No. 2, pp. 157–95.

Stevenson, G.G. (1989), 'The production, distribution, and consumption of fuelwood in Haiti', *The Journal of Developing Areas*, Vol. 24, No. 10, pp. 59–76.

Szanton, M.C.B. (1972), *A Right to Survive: Subsistence Market in a Lowland Philippine Town*, The Pennsylvania State University Press: University Park.

Teplitz-Sembitzky, W. and Schramm, G. (1989), *Woodfuel Supply and Environmental Management*, Energy Series Paper No. 19, World Bank: Washington, D.C.

Timmer, C.P., Falcon, W.P. and Pearson, S.R. (1983), *Food Policy Analysis*, The Johns Hopkins Press for The World Bank: Baltimore.

Tinker, I. (1987), 'The real rural energy crisis: women's time', *The Energy Journal*, Vol. 8, pp. 125–46

TISCO (Tanzania Industrial Studies and Consulting Organization) (1986), *Production and Market of Charcoal in Dar es Salaam*, TISCO Project 86–012: Dar es Salaam.

Trager, L. (1988), 'Customers and creditors: variations in economic personalism in a Nigerian market system', *Ethnology*, Vol. 20, No. 2, pp. 133–46.

Victus, A.M. (1993), 'Industrial Energy Use in Urban Tanzania', Working Paper, Stockholm Environment Institute: Stockholm.

Whitworth, A. (1982), 'Price control techniques in poor countries: the Tanzanian case', *World Development*, Vol. 10, No. 6, pp. 475–88.

Williamson, O.E. (1975), *Markets and Hierarchies: Analysis and Antitrust Implications*, The Free Press: New York.

World Bank (1984), *Tanzania: Issues and Options in the Energy Sector*, World Bank: Washington, D.C.

World Bank (1987a), *Review of Household Energy Options in Africa*, World Bank: Washington, D.C.

World Bank (1987b), *Tanzania Urban Woodfuels Supply Study*, World Bank: Washington, D.C.

World Bank (1992), *The World Development Report 1992*, World Bank: Washington, D.C.

Wolf, C., Jr. (1988), *Markets or Governments: Choosing Between Imperfect Alternatives*. MIT Press: Cambridge.

Wortman, S. and Cummings R.W., Jr (1978), *To Feed This World*, The Johns Hopkins University Press: Baltimore.

Appendix

The survey instruments used in the gathering of data for this project are presented in the following order:

1. Survey of end users — 'Urban Mkaa and Kuni Buyers Survey'
2. Survey of traders — 'Urban Sellers Survey'
3. Survey of transporters — 'Woodfuel Transportation Survey'
4. Survey of producers and harvesters — 'Mkaa and Kuni Producers / Cutters Survey'

Some of the *Kiswahili* words found in these surveys:
- *debe* = measure of about 2 kg used in charcoal sales.
- *fungu* = measure of about 2.5 kg used in charcoal sales in Mbeya.
- *gunia* = burlap bag used for charcoal holding 40 to 50 kg of charcoal.
- *kopo* = measure of about 1 kg used in charcoal sales.
- *kuni* = firewood.
- *mkaa* = charcoal.

TANZANIA
URBAN MKAA AND KUNI BUYERS SURVEY
INTERVIEW CONTROL / IDENTIFICATION

1. Name of interviewer: _____
2. Date of interview: _____
3. Time of interview: Begin _____ End _____
4. Duration of interview: _____ minutes
5. Name of supervisor: _____
Quality Checks: enumerator check: (signature) _____
                supervisor check: (signature) _____
                office mgmt. check: (signature) _____
==

TO BE COMPLETED BY ENUMERATOR:
Type of fuel used: _____(0.0)

Appendix 175

Name of person being interviewed_____
==
1. List all of your sources of mkaa or kuni and what portion of your supply is gotten there (be specific!!)

NAME   LOCATION   TYPE   PORTION
                                OF SUPPLY   OF SUPPLY

_____ |__|(1.0) |__|__|__|(1.1)
_____ |__|(1.2) |__|__|__|(1.3)
_____ |__|(1.4) |__|__|__|(1.5)
_____ |__|(1.6) |__|__|__|(1.7)
_____ |__|(1.8) |__|__|__|(1.9)
_____ |__|(1.a) |__|__|__|(1.b)

TYPE OF SUPPLY   1. local seller at selling point   2. mobile seller
3. seller at selling point not in walking distance from home
4. arranged delivery by wholesaler   5. rural roadside
6. rural producer (non-family)   7. rural relatives
8. collect yourself   9. cut from your own trees
10. other (specify)_____(1.c)

ENUMERATOR: CHOOSE ONE OF THE ABOVE SUPPLIERS AS DESCRIBED AND MARK IT ABOVE. ALL FOLLOWING QUESTIONS WILL REFER TO THIS SUPPLIER.

2. Why do you buy mkaa or kuni from this supplier or type of supplier?
   (0. disagree   1. agree   2. most important reason)
   a. most convenient to my home |__|(2.0)
   b. most convenient to my work |__|(2.1)
   c. cheapest prices |__|(2.2)
   d. highest quality fuel |__|(2.3)
   e. largest quantity for the price |__|(2.4)
   f. relative |__|(2.5)
   g. other seller has no fuel |__|(2.6)
   h. gathered/cut/produced myself |__|(2.7)
   i. other (specify)_____(2.9) |__|(2.8)

3. How did you get the mkaa or kuni to your home from this supplier?
   1. carry it |__|(3.0)
   2. own transport
   3. hire transport
   4. delivered by supplier

If not carried, what type of transport is used? |__|(3.1)
   TYPE OF TRANSPORT   1. lorry(_____tonnes)$_{(3.2)}$
   2. pick-up/van(_____tonnes)$_{(3.3)}$

3. animal cart 4. bicycle 5. wheelbarrow 6. headload 7. push cart
8. tractor with trailer (size_____(3.4) 9. other(specify)_____(3.5)

4. How much do you pay for mkaa or kuni from this seller?

UNIT

THIS DRY SEASON	Tsh|_|_|_| per |_|	(4.0)(4.1)
	Tsh|_|_|_| per |_|	(4.2)(4.3)
WET SEASON	Tsh|_|_|_| per |_|	(4.4)(4.5)
	Tsh|_|_|_| per |_|	(4.6)(4.7)
LAST DRY SEASON	Tsh|_|_|_| per |_|	(4.8)(4.9)
	Tsh|_|_|_| per |_|	(4.a)(4.b)

UNIT: 1.large bag 2. small bag 3. kopo 4. debe 5. log 6. large piece
7.small piece 8.bundle 9.kg 10.tonne 11.other (specify)_____(4.c)

5. Are there any differences between sellers in your area? (0.no 1.yes)

|_|(5.0)

If yes, how do they differ?   (0. no 1.yes)
    a. price |_|(5.1)
    b. quality of fuel |_|(5.2)
    c. quantity of fuel |_|(5.3)
    d. reliablity of supply |_|(5.4)

6. Do you ever give money in advance to your supplier? (0.no 1.yes)

|_|(6.0)

Do your suppliers ever give you credit?   (0.no 1.yes) |_|(6.1)

TANZANIA
URBAN SELLERS SURVEY
INTERVIEW CONTROL / IDENTIFICATION

1. Name of interviewer: _____
2. Date of interview: _____
3. Time of interview: Begin _____ End _____
4. Duration of interview: _____ minutes
5. Name of supervisor : _____
Quality Checks: enumerator check: (signature) _____
                supervisor check: (signature) _____
                office mgmt. check: (signature) _____

==

TO BE COMPLETED BY ENUMERATOR:
Type of fuel used_____(0.0)
Distance from last node:_____ (0.1)

Type of trader: (0. stationary 1. mobile) |__|(0.2)
STATIONARY District_____
Location of selling point _____
        Type of selling point: |__|(0.3)
        1. store
        2. market (#of other sellers_____(0.4)
        3. home
        4. other (specify)_____(0.5)
MOBILE Type of vehicle used |__|$_{(0.6)}$
  TYPE OF VEHICLE
  1. lorry(___tonne $_{(0.7)}$) 2. pick-up/van (___tonne $_{(0.8)}$) 3. tractor with trailer
  (size_____$_{(0.9)}$) 4. animal cart 5. bicycle 6. wheelbarrow 7. headload
  8. pushcart 10. train 11. boat 12. other (specify)_____$_{(0.a)}$

Does he work for a wholesaler, or is he an independent operator? (If he works for a wholesaler as a driver, you should interview the wholesaler instead).

Name of person being interviewed_____
Is this trader the owner of the business? _____ If not, why are you interviewing him?_____
===
1. Sex of owner: 1. male   2. female |__|$_{(1.0)}$
2. Age of owner: |__|__|years $_{(2.0)}$
3. Marital status of owner: |__|$_{(3.0)}$
   1. single 2. married 3. divorced 4. widow/widower
4. Level of school attained: |__|$_{(4.0)}$
   1. no schooling 2. some primary school 3. completed primary school
   4. secondary school 5. A-levels   6. college   7. no response
5. When did you (the owner) begin your involvement in the woodfuel business?   DATE month|__|__| year |__|__|__|__|
ENUMERATOR:CALCULATE MONTHS IN BUSINESS |__|__|__|$_{(5.0)}$
6. What is your primary economic activity? |__|$_{(6.0)}$
   1. mkaa or kuni seller
   2. farmer/agricultural worker
   3. government worker
   4. trader   (other products)
   5. other (specify)_____(6.1)
What was your primary occupation before you began selling mkaa or kuni?
                                  |__|$_{(6.2)}$
    1. student

2. farmer (own land)
3. agricultural worker
4. housework (paid)
5. housewife/mother
6. casual worker in town
7. trader (not mkaa or kuni)
8. other (specify)_____(6.3)

7. What is the primary reason that you are in this business? |__|(7.0)
    1. it is more profitable than other work
    2. can look after house/children while selling
    3. can do other economic activities while selling
    4. retired from regular job
    5. extra income on top of regular work
    6. other (specify)_____(7.1)

8. How many hours per week do you operate your woodfuel business?
    From _____ to _____ , ___ days per week |__|__|(8.0)

9. How many months in the year do you sell woodfuel? |__|__|(9.0)

10. If you are a stationary seller, why did you pick this location for selling your woodfuel?
                    (0. no 1. yes   2. most important reason)
    a. near to place where you buy woodfuel (wholesale outlet) |__|(10.0)
    b. convenient to/at home |__|(10.1)
    c. not too many other woodfuel sellers |__|(10.2)
    d. near to many customers or potential customers |__|(10.3)

Are there any other reasons?_____

11. Do you ever get paid money in advance for your mkaa or kuni?
                                      (0. no 1. yes) |__|(11.0)

If yes, how much do you get paid? Tsh |__|__|__|__|__|(11.1)
how far in advance? |__|__|(11.2)_____ (specify day or month or year)
at what price do you give them the mkaa or kuni?
                        Tsh |__|__|__|__|__| per |__| (11.3)(11.4)
UNIT: 1. large bag 2. small bag 3. kopo 4. debe 5. log 6. large piece 7. small piece
       8. bundle 9. kg 10. tonne 11. other (specify)_____(11.5)

12. Do you ever give credit?   (0. no 1. yes) |__|(12.0)

13. Have you taken any loan to use for your business? |__|(13.0)
                                                      (0. no 1. yes)

14. Labour force and salaries or allowances paid per month or job

Non-family workers

NUMBER	JOB TYPE(S)	PAYMENT (Tsh) (PER WORKER)	WHEN PAID?		
full-time					
|__|__|	|__|	|__|__|__|__|__|	|__|		
|__|__|	|__|	|__|__|__|__|__|	|__|		
|__|__|	|__|	|__|__|__|__|__|	|__|		
part-time/occasional			HOW OFTEN?		TIME
|__|__|	|__|	|__|__|__|__|__|	|__|	|__|__|times per	|__|
|__|__|	|__|	|__|__|__|__|__|	|__|	|__|__|times per	|__|
|__|__|	|__|	|__|__|__|__|__|	|__|	|__|__|times per	|__|

family workers(including owner)

	NUMBER	JOB TYPE(S)
full-time	|__|__|	|__|
	|__|__|	|__|
	|__|__|	|__|
part-time/ occasional	|__|__|	|__|
	|__|__|	|__|
	|__|__|	|__|

WHEN PAID   1. weekly 2. monthly 3. every 2 weeks 4. Daily
5. after job is completed 6. before starting job 7. after mkaa/kuni is sold
TIME   2.day 3.week 4..month   5.year
JOB TYPE 1. load/unload vehicle with mkaa or kuni   2. split firewood into smaller pieces 3. sell woodfuel at market   4. deliver woodfuel to customers.
5. transport woodfuel from wholesaler to selling point 6. askari   7. other (specify)_____8. other (specify)_____ 9. other (specify)_____

Demand

15. Indicate the normal daily volume of sales.

UNIT	WET SEASON	DRY SEASON
|__|(15.0)	|__|__|__|(15.1)	|__|__|__|(15.2)
|__|(15.3)	|__|__|__|(15.4)	|__|__|__|(15.5)
|__|(15.6)	|__|__|__|(15.7)	|__|__|__|(15.8)
|__|(15.9)	|__|__|__|(15.a)	|__|__|__|(15.b)

16. What portion or percentage of your sales are to regular customers and what percentage to other customers?

	%, PORTION or #of bags
a. regular customers	|__|__|__|(16.0)
b. other customers	|__|__|__|(16.1)

17. What portion or percentage of your sales are to each type of customer?
    %, PORTION or #of bags
    a. households |__|__|__|(17.0)
    b. businesses |__|__|__|(17.1)
    c. other mkaa or kuni sellers |__|__|__|(17.2)
    d. transporters |__|__|__|(17.3)
18. At what price do you sell your mkaa or kuni?
WET SEASON
    UNIT   PRICE AT SELLING POINT   PRICE IF DELIVERED
    |__|(18.0)   Tsh |__|__|__|__|(18.1)   Tsh |__|__|__|__|(18.2)
    |__|(18.3)   Tsh |__|__|__|__|(18.4)   Tsh |__|__|__|__|(18.5)
DRY SEASON
    UNIT   PRICE AT SELLING POINT   PRICE IF DELIVERED
    |__|(18.6)   Tsh |__|__|__|__|(18.7)   Tsh |__|__|__|__|(18.8)
    |__|(18.9)   Tsh |__|__|__|__|(18.a)   Tsh |__|__|__|__|(18.b)
LAST DRY SEASON
    UNIT   PRICE AT SELLING POINT   PRICE IF DELIVERED
    |__|(18.c)   Tsh |__|__|__|__|(18.d)   Tsh |__|__|__|__|(18.e)
    |__|(18.f)   Tsh |__|__|__|__|(18.g)   Tsh |__|__|__|__|(18.h)
19. How do you set your prices? |__|(19.0)
    1. negotiation with buyers
    2. talking to other traders and setting a price together
    3. matching the price of other traders
    4. taking all your costs plus a profit margin
    5. price of mkaa or kuni bought plus a profit margin
    6. by comparing with prices of other fuels
    7. other (specify)_____(19.1)
20. How often do you change your prices |__|(20.0)
    1. once a year in wet season   2. more often
    What causes you to change them? |__|(20.1)
    1. transporters raise prices because of season
    2. transporters raise prices because of fuel costs
    3. producers/cutters raise prices
    4. wholesalers raise price
21. Do you ever set your prices below those of other traders? |__|(21.0)
                                               (0. no 1.yes)
Why or why not?_____
22. What is the type of mkaa or kuni preferred by your customers?
        1. _____(22.0)

    2. _____ (22.1)
    3. _____ (22.2)
    4. no preference |__|(22.3)
    9. do not know

23. Do you select the mkaa or kuni you buy? (0. no 1. yes) |__|(23.0)
    Do you reject low quality fuel?   (0. no 1. yes) |__|(23.1)
    If you have low quality fuel to sell, do you: (0. no 1. yes)
        1. sell at a discount |__|(23.2)
        2. give more for the same price |__|(23.3)
    Does it take you longer to sell low quality mkaa or kuni? |__|(23.4)

Supply

23.5. At what price do you buy your mkaa or kuni?

WET SEASON
    UNIT   PRICE AT SELLING POINT   PRICE IF DELIVERED
    |__|(18.0)   Tsh |__|__|__|__|(18.1)   Tsh |__|__|__|__|(18.2)
    |__|(18.3)   Tsh |__|__|__|__|(18.4)   Tsh |__|__|__|__|(18.5)

DRY SEASON
    UNIT   PRICE AT SELLING POINT   PRICE IF DELIVERED
    |__|(18.6)   Tsh |__|__|__|__|(18.7)   Tsh |__|__|__|__|(18.8)
    |__|(18.9)   Tsh |__|__|__|__|(18.a)   Tsh |__|__|__|__|(18.b)

LAST DRY SEASON
    UNIT   PRICE AT SELLING POINT   PRICE IF DELIVERED
    |__|(18.c)   Tsh |__|__|__|__|(18.d)   Tsh |__|__|__|__|(18.e)
    |__|(18.f)   Tsh |__|__|__|__|(18.g)   Tsh |__|__|__|__|(18.h)

24. Indicate your main suppliers of mkaa or kuni
NAMES, LOCATIONS (BE SPECIFIC), PORTION SUPPLIED
                                        TYPE   PORTION
_____ |__|(24.1) |__|__|__|(24.2)
_____(24.0)
_____ |__|(24.4) |__|__|__|(24.5)
_____(24.3)
_____ |__|(24.7) |__|__|__|(24.8)
_____(24.6)

    TYPE: 1. transporter 2. stationary selling point   3. rural wholesaler
    4. rural producer/cutter

ENUMERATOR: CHOOSE ONE SUPPLIER AS DIRECTED, AND MARK IT ABOVE. BE SURE THAT YOU HAVE ENOUGH INFORMATION ABOUT THE SELECTED SUPPLIER THAT YOU CAN FIND HIM LATER. QUESTIONS 25 THROUGH 28 WILL REFER TO

THIS SUPPLIER OR TYPE OF SUPPLIER. TELL THE RESPONDENT THAT WE WILL BE ASKING HIM QUESTIONS ABOUT THAT SUPPLIER.

Supplier Specific Questions
25. Is this supplier a producer of mkaa or a cutter/collector of kuni?
                                      (0. no 1. yes) |__|$_{(25.0)}$
    Is he a relative of yours?   (0. no 1. yes) |__|$_{(25.1)}$
    If no, is he from the same village as you? (0. no 1. yes) |__|$_{(25.2)}$
26. How is mkaa/kuni from this supplier transported and delivered to you?
                                                                   |__|$_{(26.0)}$
    1. you pick it up from a seller in town 2. you pick it up from rural area
    3. transport hired by you   4. delivered by supplier
    5. other (specify)_____(26.1)
(IF 2, ALSO GIVE TRANSPORT SURVEY)
What type of transport is most used to deliver the fuel? |__|$_{(26.2)}$
    TYPE OF TRANSPORT   1. lorry(___tonne$_{(26.3)}$)
    2. pick-up/van (___tonne$_{(26.4)}$) 3. tractor with trailer (_____$_{(26.5)}$)
    4. animal cart 5. bicycle 6. wheelbarrow 7. headload 8. pushcart 10. train
    11. boat 12. other (specify)_____(26.6)
27. If you hire transport, how much do you pay for transport per load?
                         AMOUNT
                Tsh |__|__|__|__|__|per load $_{(27.0)}$
        Size of load |__|__|__| |__|(UNIT)$_{(27.1)}$
    UNIT: 1.large bag 2. small bag 3. kopo 4. debe 5. log 6. large piece
    7.small piece 8.bundle 9.kg 10.tonne 11.other (specify)_____(27.2)
Does the driver usually carry cargo out of the city and then carry your mkaa or kuni back into the city? |__|$_{(27.3)}$ (0. never 1. usually 2. always)
Do you always use the same transporter?   (0. no 1.yes) |__|$_{(27.4)}$
28. Do you have an agreement with this supplier: 0. no 1. yes
    to always buy his mkaa or kuni? |__|$_{(28.0)}$
    that he should supply mkaa or kuni on a regular basis? |__|$_{(28.1)}$

General Questions
29. Do you buy gunia for mkaa?   (0. no 1.yes) |__|$_{(29.0)}$
If yes, how much do you pay for them?
                Tsh |__|__|__|__| per |__|__| bags $_{(29.1)(29.2)}$
How many times can you use each gunia? |__|__| times $_{(29.3)}$
    Are they difficult to get? (0. no 1.yes) |__|$_{(29.4)}$

30. How frequently do you get mkaa or kuni?
     wet season |__|__|times/|__|__|days (30.0)(30.1)
     dry season |__|__|times/|__|__|days (30.2)(30.3)
How much mkaa or kuni do you buy each time?
     |__|__|__|(30.4)|__|(UNIT)(30.5)
UNIT: 1.large bag 2. small bag 3. kopo 4. debe 5. log 6. large piece
7.small piece 8.bundle 9 kg 10.tonne 11.other (specify)_____(30.6)

31. Is it more difficult to get mkaa or kuni in one season or another?
     1. more difficult in rainy season |__|(31.0)
     2. more difficult in dry season
     3. no difference between the two seasons

32. If you sell from a selling point, how much do you pay for the use of that location for your business?   Tsh |__|__|__|__| per month (32.0)

33. If you are a mobile seller, how far do you travel in a day of selling?
     |__|__|__| km (33.0)

34. Do you ever buy mkaa or kuni in the dry season and store it to sell in the wet season?   (0. no   1.yes) |__|(34.0)
Why or why not?_____

---

35. How much mkaa or kuni do you have at this time?
                    NUMBER   UNIT
                    |__|__|__|(35.0) |__|(35.1)
                    |__|__|__|(35.2) |__|(35.3)
Where is it stored? |__|(35.4)
1. warehouse 2. Residence 3. store (selling point) 4. other (specify)_____(35.5)
How far is that from where you sell it? |__|__|__| km (35.6)
How far it that from where you buy it? |__|__|__| km (35.7)
Does it cost you anything extra to store it? Tsh|__|__|__|per month(35.8)
What is the maximum you store at any time?
     |__|__|__|(35.9)|__|(UNIT)(35.a)
UNIT: 1.large bag 2. small bag 3. kopo 4. debe 5. log 6. large piece
7.small piece 8.bundle 9.kg 10.tonne 11.other(specify)_____(35.b)

Perceptions
36. Is it harder to get mkaa or kuni now than it used to be? |__|(36.0)
Is it harder to find buyers for mkaa/kuni now than it used to be? |__|(36.1)
37. Have the supply areas changed in the past 5 years? |__|(36.0)
     0. no   1. yes   9. don't know
If yes indicate the 3 main former areas of supply:

Area 1 _____ (37.0)
Area 2 _____ (37.1)
Area 3 _____ (37.2)

38. In your opinion who is making the largest profit in the mkaa or kuni business? |__|(38.0)
    1. transporters 2. wholesalers   3. producers
    4. government   5. retailers   6. other (specify)_____(38.1)

39. What are the major problems in your business?
    0. do not agree 1. agree somewhat   2. agree strongly
    a. too many mkaa or kuni sellers |__|(39.0)
    b. wholesalers and transporters charge too much |__|(39.1)
    c. difficult to obtain loans |__|(39.2)
    d. difficult to get supply of mkaa or kuni |__|(39.3)
    e. not enough profit |__|(39.4)
    f. fees and licensing too expensive |__|(39.5)
    g. difficult to get supplies of bags for charcoal |__|(39.6)
    h. other (specify)_____(39.8) |__|(39.7)

40. In your opinion, what can be done to solve these problems?

---

41. Which of the following fees or taxes do you pay?   0. no 1. yes
    a. sales tax |__|(41.0)   Tshl__|__|__|
        perl__|(UNIT)(41.1)(41.2)
    b. government taxes on mkaa or kuni |__|(41.3)
        Tshl__|__|__| perl__|(UNIT)(41.4)(41.5)
    c. local taxes on mkaa or kuni |__|(41.6)
        Tshl__|__|__| perl__|(UNIT)(41.7)(41.8)
    d. other (specify)_____(41.a) |__|(41.9)
        Tshl__|__|__| perl__|(UNIT)(41.b)(41.c)
    UNIT: 1.large bag 2. small bag 3. kopo 4. debe 5. log 6. large piece
    7.small piece 8.bundle 9.kg 10.tonne 11.other(specify)_____(41.d)

42. WEIGHT OF DIFFERENT UNITS (kg)

UNITS	1	2	3	4	5	AVG.	
_____ (42.1)	___	___	___	___	___	___	|__|(42.0)
_____ (42.3)	___	___	___	___	___	___	|__|(42.2)
_____ (42.5)	___	___	___	___	___	___	|__|(42.4)

___ ___ ___ ___ ___ ___ ___ |__|(42.6)
(42.7)

UNIT: 1.large bag 2. small bag 3. kopo 4. debe 5. log 6. large piece
7.small piece 8.bundle 9.kg 10.tonne 11.other (specify)_____ (42.8)

TANZANIA
WOODFUEL TRANSPORTATION SURVEY
INTERVIEW CONTROL / IDENTIFICATION

1. Name of interviewer: _____
2. Date of interview: _____
3. Time of interview: Begin _____ End _____
4. Duration of interview: _____ minutes
5. Name of supervisor: _____
Quality Checks: enumerator check: (signature) _____
               supervisor check: (signature) _____
               office mgmt. check: (signature) _____

==

TYPE OF FUEL_____(0.0)
Name of transporter: _____
Address:_____
++

1. Sex: 1. male 2. female |__|(1.0)
2. Age: |__|__|years(2.0)
3. What is your *primary* economic activity? |__|(3.0)
    1. mkaa or kuni transporter
    2. transporter of other goods - self employed
    3. transporter of other goods for employer
    4. driver for private company or individual
    5. driver for government or parastatal
    6. other (specify)_____(3.1)
4. When did you begin your involvement in the woodfuel business?
        DATE   month|__|__| year |__|__|__|__|
ENUMERATOR:CALCULATE MONTHS IN BUSINESS |__|__|__|(4.0)
5. What was your primary occupation before that? |__|(5.0)
    1. student
    2. transporter (other products)
    3. driver
    4. farmer (own land)
    5. agricultural worker

      6. casual worker in town
      7. government worker
      8. other (specify)_____(5.1)

6. What is the primary reason you are in this business? |__|(6.0)
      1. can make more money than carrying other goods
      2. family business
      3. extra income on top of regular job
      4. other (specify)_____(6.1)

7. Who is the owner of the vehicle? |__|(7.0)
  1. self   2. company   3. relative   4. other (specify)_____(7.1)

8. If you own the vehicle, have you taken any loan(s) to use for your business?   (0. no   1. yes) |__|(8.0)

9. How old and in what condition were the vehicles that you use for woodfuel transportation when they were acquired?

TYPE   YEAR BOUGHT   YEAR OF   CONDITION AT
                       MANUFACTURE   TIME OF PURCHASE
|__|$_{(9.0)}$|__|__|__|__|$_{(9.1)}$ |__|__|__|__|$_{(9.2)}$ |__|$_{(9.3)}$

TYPE OF VEHICLE: 1.lorry (____tonne$_{(9.4)}$)   2. pick-up/van(____tonne$_{(9.5)}$) 3. tractor with trailer   4. other (specify)_____(9.6)

CONDITION: 1. new   2. secondhand   3. scrap   4. don't know

10. Are you ever hired to carry mkaa or kuni for other people? |__|(10.0)
                                                         (0. no   1.yes)

    If yes, who hires you? |__|(10.1)
    1. wholesalers in town   2. wholesalers in rural areas
    3. other (specify)_____(10.2)

How much are you paid? Tsh |__|__|__|__|__| per |__|(UNIT)$_{(10.3)(10.4)}$
    UNIT: 1.large bag 2. small bag 3. kopo 4. debe 5. log 6. large piece 7.small piece   8.bundle   9.kg   10.tonne   11.other (specify)_____(10.5)^3

SUPPLY

11. Where do you buy mkaa or kuni?
LOCATION (be as specific as possible! Include village name, producer name if known, forest reserve name if applicable, etc.)

	TYPE OF LAND	TYPE OF PRODUCER	PROPORTION
1. _____ |__|(11.0)	|__|(11.1)	|__|__|__|(11.2)
_____(11.3)			
2. _____ |__|(11.4)	|__|(11.5)	|__|__|__|(11.6)
_____(11.7)			
3. _____ |__|(11.8)	|__|(11.9)	|__|__|__|(11.a)

─────────────────────────────(11.b)
TYPE OF PRODUCER 1. fulltime producer of mkaa at site 2.one-time producer of mkaa 3. roadside seller 4.fulltime cutter or collector of kuni, not at roadside 5.one time cutter or collector of kuni

TYPE OF LAND 1.forest reserve 2.public land 3.own farm land

ENUMERATOR: CHOOSE ONE OF THESE SUPPLIERS/SUPPLY AREAS AND MARK IT ABOVE. BE SURE THAT YOU HAVE ENOUGH INFORMATION ABOUT LOCATION THAT YOU CAN FIND THE SUPPLIER YOURSELF. REFER TO THIS SUPPLIER/SUPPLY AREA FOR QUESTIONS 12 THROUGH 14.

Supplier Specific Questions

12. At what price do you buy mkaa or kuni?
DRY SEASON   Tsh|__|__|__|__|per|__|$_{(12.0)(12.1)}$
WET SEASON   Tsh|__|__|__|__|per|__|$_{(12.2)(12.3)}$
LAST DRY SEASON Tsh|__|__|__|__|per|__|$_{(12.4)(12.5)}$

UNIT: 1.large bag 2. small bag 3. kopo 4. debe 5. log 6. large piece 7.small piece 8.bundle 9.kg 10.tonne 11.other (specify)_____$_{(12.6)}$

13. How long do you usually take to go and return from this supplier?
    DRY SEASON |__|__|__|.|__|__| days $_{(13.0)}$
    WET SEASON |__|__|__|.|__|__| days $_{(13.1)}$

How far is it from the city (DSM, Mbeya, or Shinyanga, as appropriate)?
|__|__|__| km $_{(13.2)}$

If you buy from the site, how far is it from the main road?|__|__|km $_{(13.3)}$

14. Is this supplier related to you?   (0. no 1.yes) |__|$_{(14.0)}$
Is the supplier from the same village as you? (0. no 1.yes) |__|$_{(14.1)}$

General Questions

15. Are your supply areas different in the wet season? (0.no 1.yes)|__|$_{(15.0)}$
    If yes, where are they (be specific, as before)

	TYPE OF LAND	TYPE OF PRODUCER	PROPORTION
1._____	|__|$_{(15.0)}$	|__|$_{(15.1)}$	|__|__|__|$_{(15.2)}$
_____(15.3)			
2._____	|__|$_{(15.4)}$	|__|$_{(15.5)}$	|__|__|__|$_{(15.6)}$
_____(15.7)			
3._____	|__|$_{(15.8)}$	|__|$_{(15.9)}$	|__|__|__|$_{(15.a)}$

─────────────────────────────(15.b)
TYPE OF PRODUCER: 1. Full time producer of mkaa at site 2.one-time producer of mkaa 3. roadside seller 4.full time cutter or collector of kuni,

not at roadside   5.one time cutter or collector of kuni
TYPE OF LAND   1. forest reserve   2. public land 3. own farm land

Why do you change to these suppliers in the wet season? |__|(15.c)
   1. Unable to reach dry season suppliers because of road conditions
   2. Dry season suppliers don't have enough mkaa or kuni in wet season
   3. Prefer them, but don't supply enough during dry season so must use other suppliers then   4. other (specify)_____(15.d)

If no, are you ever prevented from reaching a supplier because of road conditions?   (0. no 1.yes) |__|(15.e)

16. How often do you transport mkaa or kuni into town?
   Dry season |__|__|trips in |__|__|days $(16.0)(16.1)$
   Wet season |__|__|trips in |__|__|days $(16.2)(16.3)$

17. How many or which months are in the wet season?_____(17.0)

18. How much mkaa or kuni do you transport per trip?
                |__|__|__| |__|(UNIT)$(18.0)(18.1)$
UNIT: 1.large bag 2. small bag 3. kopo 4. debe 5. log 6. large piece
7.small piece 8.bundle 9.kg 10.tonne 11.other (specify)_____(18.2)

19. During which part of the day or night do you do most of your mkaa or kuni transportation? |__|(19.0)
      1. early in the morning (up to noon)
      2. in the afternoon (up to 18:00)
      3. anytime during daylight
      4. at night

Why do you transport mkaa or kuni at that time?
      (0. no 1.yes 2. most important reason)
   a. less traffic |__|(19.1)
   b. fewer road blocks |__|(19.2)
   c. have other work during the night/day |__|(19.3)
   d. cooler |__|(19.4)
   e. get to town at best time to sell mkaa or kuni |__|(19.5)
   f. safest time to drive |__|(19.6)
   g. other (specify) _____ (19.8) |__|(19.7)

20. What portion of trips do you transport:
a. empty on trip out of town, mkaa or kuni only on return |__|__|__|(20.0)
b. empty on trip out of town, mkaa or kuni & cargo on return|__|__|__|(20.1)
c. cargo on trip out of town, mkaa or kuni only on return |__|__|__|(20.2)
d. cargo on trip out of town, mkaa or kuni & cargo on return|__|__|__|(20.3)
e. vehicle used for other things on trip out of town, return with mkaa or kuni |__|__|__|(20.4)

21. What portion of your supplies are gotten from suppliers you buy

from regularly? |__|__|__|(21.0)
22. Do you ever give money in advance to your suppliers? |__|(22.0)
(0. no 1.yes)
    If yes, do you then get the mkaa or kuni at a discount?|__|(22.1)
(0. no 1.yes)
    Do your suppliers ever sell to you an credit? (0. no 1.yes) |__|(22.2)
23. Do you transport mkaa or kuni to other towns? (0. no 1. yes) |__|(23.0)
    If so, indicate the towns:
1._____(23.1) 2._____(23.2)
24. Do you provide gunia to the mkaa suppliers? (0. no 1.yes) |__|(24.0)
If yes, how much do you pay for them?
                Tsh |__|__|__|__| per |__|__| gunia (24.0)(24.1)
How often do you have to replace them? Every |__|__| weeks (24.2)
    How many times can you use one gunia? |__|__| times (24.3)
25. Labour force and salaries or allowances paid per month or job
Non-family workers

NUMBER	JOB TYPE(S)	PAYMENT (Tsh) (PER WORKER)	WHEN PAID?		
full-time					
|__|__|	|__|	|__|__|__|__|__|	|__|		
|__|__|	|__|	|__|__|__|__|__|	|__|		
|__|__|	|__|	|__|__|__|__|__|	|__|		
part-time/occasional				HOW OFTEN?	TIME
|__|__|	|__|	|__|__|__|__|__|	|__|	|__|__|times per	|__|
|__|__|	|__|	|__|__|__|__|__|	|__|	|__|__|times per	|__|
|__|__|	|__|	|__|__|__|__|__|	|__|	|__|__|times per	|__|

Family workers   (including owner)

	NUMBER	JOB TYPE(S)
full-time	|__|__|	|__|
	|__|__|	|__|
	|__|__|	|__|
part-time/	|__|__|	|__|
occasional	|__|__|	|__|
	|__|__|	|__|

       WHEN PAID   1. weekly 2. Monthly 3. every 2 weeks 4. Daily 5. after job is completed 6. before starting job 7.after mkaa/kuni is sold
       TIME   2.day 3.week 4..month 5.year

JOB TYPE 1. load/unload vehicle with mkaa or kuni   2. split firewood into smaller pieces   3. fill sacks with mkaa   4. driver   5. maintain vehicle   6. Askari   7. other (specify)_____   8. other (specify)_____   9. other (specify)_____

DEMAND
26. At what price do you sell mkaa or kuni?
DRY SEASON   Tsh|__|__|__|__|per|__|$_{(26.0)(26.1)}$
WET SEASON   Tsh|__|__|__|__|per|__|$_{(26.2)(26.3)}$
LAST DRY SEASON   Tsh|__|__|__|__|per|__|$_{(26.4)(26.5)}$
UNIT: 1.large bag 2. small bag 3. kopo 4. debe 5. log 6. large piece
7.small piece 8.bundle 9.kg 10.tonne 11.other (specify)_____$_{(26.6)}$
27. How much mkaa or kuni do you sell in a normal week?
    dry season |__|__|__| |__|(UNIT)$_{(27.0)(27.1)}$
    wet season |__|__|__| |__|(UNIT)$_{(27.2)(27.3)}$
28. What are the maximum and minimum amount of mkaa or kuni you usually sell to any one buyer?
    maximum |__|__|__| |__|(UNIT)$_{(28.0)(28.1)}$
    minimum |__|__|__| |__|(UNIT)$_{(28.2)(28.3)}$
29. How long does it normally take you to sell one load of mkaa or kuni <u>after</u> you have reached town?
    dry season |__|__|.|__|__| days $_{(29.0)}$
    wet season |__|__|.|__|__| days $_{(29.1)}$
If it takes longer in one season or the other, why is this? |__|$_{(29.2)}$
    1. fewer people buying mkaa or kuni
    2. more transporters selling mkaa or kuni
    3. difficult to get to buyers
    4. other (specify)_____$_{(29.3)}$
How many litters of fuel (or how many Tsh worth) do you use selling it?
|__|__|__| liters$_{(29.4)}$ (OR Tsh|__|__|__|__|) of|__|(TYPE OF FUEL)$_{(29.5)}$
TYPE OF FUEL: 1. diesel 2. petrol 3. kerosene
30. If you transport <u>only</u> mkaa or kuni, how many litters of fuel (or how many Tsh worth) do you use going to the rural supplier and back to town <u>and</u> selling the mkaa or kuni in town?
|__|__|__| liters$_{(30.1)}$ (OR Tsh|__|__|__|__|) of|__|(TYPE OF FUEL)$_{(30.2)}$
TYPE OF FUEL: 1. diesel 2. petrol 3. kerosene

31. What portion of your buyers are regular buyers |__|__|__|$_{(31.0)}$
32. Do you ever give credit to your buyers? (0. no 1.yes) |__|$_{(32.0)}$

PERCEPTIONS

33. What changes have taken place in the mkaa or kuni transportation business during the last 5 years? |__|(33.0)
1. decline in business 2. no change in business 3. growth in business

34. Who do you think makes the most profit in the mkaa or kuni business?
|__|(34.0)
1. producers 2. wholesalers 3. retailers
4. transporters 5. government 6. other (specify)_____(34.1)

35. The major problems in the mkaa or kuni transportation business are:
(0. do not agree 1. agree 2. most important reason)
a. long distances to and from collection/production areas |__|(35.0)
b. lack of vehicle spare parts |__|(35.1)
c. expensive vehicle spare parts |__|(35.2)
d. high cost of replacement vehicles |__|(35.3)
e. high fuel and maintenance costs |__|(35.4)
f. high Labour costs |__|(35.5)
g. stiff competition from other transporters |__|(35.6)
h. lack of good roads |__|(35.7)
i. other (specify)_____(35.9) |__|(35.8)

36. In your opinion, what do you think should be done to improve the mkaa or kuni transportation business?_____(36.0)

37. Which of the following fees or taxes do you pay on mkaa or kuni?
(0.no 1.yes)

a. forest service tax |__|(37.0) Tsh|__|__|__|(37.1) per|__|(UNIT)(37.2)
b. district taxes |__|(37.3) Tsh|__|__|__|(37.4) per|__|(UNIT)(37.5)
c. city taxes |__|(37.6) Tsh|__|__|__|(37.7) per|__|(UNIT)(37.8)
d. village taxes |__|(37.9) Tsh|__|__|__|(37.a) per|__|(UNIT)(37.b)
e. other_____37.f) |__|(37.c) tsh|__|__|__|(37.d) per|__|(UNIT)(37.e)

TANZANIA
MKAA AND KUNI PRODUCERS/CUTTERS SURVEY
INTERVIEW CONTROL/ IDENTIFICATION

1. Name of interviewer: _____
2. Date of interview: _____
3. Time of interview Begin _____ End _____
4. Duration of interview:_____ minutes
5. Name of supervisor:

Quality Checks: enumerator check: (signature) _____
supervisor check: (signature) _____
office mgmt. check: (signature) _____

===

TO BE COMPLETED BY ENUMERATOR:
Type of fuel produced: _____ $_{(0.1)}$ IF MKAA, COMPLETE PART 1 AND PART 2. IF KUNI, COMPLETE PART 1 AND PART 3.
Location of interview (region, district, village, etc.)_____
Distance from pick-up point to main road_____(0.2)
Distance from city (DSM, Mbeya, or Shinyanga)_____(0.3)
Distance from kiln to pick-up point OR from cutting area to pick-up point_____(0.4)
Name of person being interviewed_____

===

PART 1 – GENERAL QUESTIONS

1. Sex: 1. male 2. female |__|$_{(1.0)}$
2. Age: |__|__|years$_{(2.0)}$
3. What is your primary economic activity? |__|$_{(3.0)}$
    1. mkaa or kuni supplier
    2. farmer (own land)
    3. farm worker
    4. other (specify)_____(3.1)
4. When did you begin making mkaa or cutting/collecting kuni to sell?
    DATE month|__|__| year |__|__|__|__|
ENUMERATOR: CALCULATE MONTHS IN BUSINESS |__|__|__|$_{(4.0)}$
5. What is the primary reason you are in this business? |__|$_{(5.0)}$
    1. extra income on top of farming
    2. family business
    3. other (specify)_____(5.1)
6. Do you come from this village or a nearby one? (0. no 1.yes) |__|$_{(6.0)}$
    If no, where do you come from? _____(6.1)
7. Do the buyers collect your mkaa or kuni from you at the site where you produced it or from the roadside? |__|$_{(7.0)}$ 1. site 2. roadside
    If it is collected at the roadside, how do you get it there? |__|$_{(7.1)}$
        1. carry it 2. own transport 3. hire transport
    How much do you pay for transport?
    Tsh|__|__|__|__|__|per|__|(UNIT)$_{(7.2)(7.3)}$
      UNIT: 1.large bag 2. small bag 3. log 4. large piece 5. small piece
      6. bundle 7. kg 8. tonne 9. other (specify)_____(7.4)

What type of transport is used? |__|(7.5)
 TYPE OF TRANSPORT: 1. lorry (____ tonnes$_{(7.6)}$) 2. animal cart 3. headload
 4. pick-up/van (____tonnes$_{(7.6)}$) 5. wheelbarrow 6. push cart 7. bicycle
 8. tractor with trailer (size____$_{(7.8)}$) 9. other (specify)_____$_{(7.6)}$
Who loads the truck?   1. you 2. buyer/employees of buyer |__|(7.7)
8. Do you always sell to the same buyer?   (0. no 1. yes) |__|(8.0)
 If no, is it ever difficult to find a buyer? (0. no 1. yes) |__|(8.1)
Are your buyers your relatives? (0. never 1. sometimes 2. always) |__|(8.2)
Does the buyer ever pay you in advance for your mkaa or kuni?
                                 (0. no 1. yes) |__|(8.3)
 If yes, do you sell your mkaa or kuni to this buyer for less than you
         would to other buyers?   (0. no 1. yes) |__|(8.4)
    If yes, what is the price to him? Tsh|__|__|__|__| per |__|(8.5)(8.6)
    UNIT: 1. large bag 2. small bag 3. kopo 4. debe 5. log 6. large piece 7. small
    piece 8. bundle 9. kg 10. tonne 11. other (specify)_____(8.7)
9. Is it more difficult to sell your mkaa or kuni in one season? |__|(9.0)
           (1. wet season 2. dry season 3. no difference)
 Why?_____(9.1)
10. Do you ever keep your mkaa or kuni until wet season in order to get a
better price for it later?   (0. no 1. yes) |__|(10.0)
   How long do you keep it? |__|__| |__|(TIME)$_{(10.1)(10.2)}$
   What is the most you ever keep? |__|__|__| |__|(UNIT)$_{(10.3)}$
   TIME 1. hour 2. day 3. week 4. month 5. year
11. Where do you usually cut trees?
               AREA   TYPE OF AREA
         _____ |__|(11.0)
         _____ |__|(11.1)
         _____ |__|(11.2)

   TYPE OF AREA 1. own land 2. land of another individual 3. village land
      4. district land 5. government forest or reserve 6. other_____(11.3)
If it is village or district land, is it controlled by your own village or
district?   (0. no 1. yes) |__|(11.4)
12. How far is it from your home? |__|__|__|.|__| km $_{(12.0)}$
    How far is it from your farm? |__|__|__|.|__| km $_{(12.1)}$
13. Do you have to ask permission to cut there? (0. no 1. yes) |__|(13.1)
    Do you have to pay anything to cut there?   (0. no 1. yes) |__|(13.2)
      If yes, how much? Tsh |__|__|__|__| per |__|(UNIT)$_{(13.3)(13.4)}$
         Who do you pay?_____(13.5)

UNIT: 1.large bag 2. small bag 3. kopo 4. debe 5. log 6. large piece
7.small piece 8.bundle 9.kg 10.tonne 11.other (specify)_____(13.6)

14. Do you cut in different areas than you did five years ago?
         (0. no 1.yes) |__|(14.0)

If yes, are they:
a. further from your farm or home |__|(14.1)
b. further from a road |__|(14.2)
c. areas with fewer of the preferred species of trees |__|(14.3)
d. areas which are now in reserves and other protected forests but weren't before. |__|(14.4)
e. areas which have been cut before and are now being recut |__|(14.5)
f. areas which are being cleared for agriculture |__|(14.6)

15. When you are cutting in an area, do you cut every tree? |__|(15.1)
(0.no 1. yes)
   If no, how do you determine which trees to cut down?
(0. no 1.yes)
         a. only trees of a certain size |__|(15.2)
         b. only trees of a certain species |__|(15.3)
         c. leave some trees to preserve the land |__|(15.4)

16. How long does it take an area that has been cut to recover so that it can be cut again? |__|__| years (16.0)

17. What types of trees do you prefer to cut?
      NAMES
    _____(17.0)
    _____(17.1)
    _____(17.2)
    _____(17.3)
    _____(17.4)
    _____(17.5)

18. Do you usually or always make mkaa or kuni from trees cut to clear land for agriculture? (0. no 1.yes) |__|(18.0)

19. Give your opinions on the following: (0. no 1.yes)
   a. Are the trees in this area getting more scarce? |__|(19.0)
   b. Are you paid a fair price for your mkaa or kuni ? |__|(19.1)
   c. Are the places where you have cut down the trees now being used for crops? |__|(19.2)

20. Do you ever plant trees to use for fuel? (0. no 1.yes) |__|(20.0)
      If yes, how many? |__|__|__| per |__| (TIME)(20.1)(20.2)
      where?   (1.On farm 2. At home 3. On village land 4. other) |__|(20.3)

If no, why not?   (0. no 1.yes 2.most important reason)
    a. already enough trees around |__|(20.4)
    b. trees grow from stumps, so no need to plant them. |__|(20.5)
    c. can not get seedlings |__|(20.6)
    d. never considered it |__|(20.7)
    e. too expensive |__|(20.8)
    f. takes too much work |__|(20.9)
    g. would not own the trees even if you planted them. |__|(20.a)
21. What fees or taxes do you pay? (0. no 1.yes)
    a. local/village tax   (Tsh|__|__|__|per|__|(UNIT)$_{(21.0)(21.1)}$ |__|$_{(21.2)}$
    b. other fee to cut trees in an area
        (Tsh|__|__|__|per |__|(UNIT)$_{(21.3)(21.4)}$)|__|$_{(21.5)}$
    c. forestry tax   (Tsh|__|__|__|per|__|(UNIT)$_{(21.6)(21.7)}$)|__|$_{(21.8)}$

==

PART 2 – MKAA

22. Do you produce mkaa regularly? (0. no 1.yes) |__|$_{(22.0)}$
  IF NO ANSWER ONLY QUESTIONS 25, 26, 30, and 31
23. How long have you been producing mkaa |__|__|.|__|__|years$_{(23.0)}$
                  OR since |__|__|__|__|(year)
24. Do you produce more charcoal in the dry season or the wet season?
(1. dry 2. wet) |__|$_{(23.0)}$ Why?_____(23.1)
25. How many kilns do you fire?   TIME
DRY SEASON |__|__|__| per |__|$_{(25.0)(22.1)}$
WET SEASON |__|__|__| |__|$_{(25.2)(22.3)}$
    TIME 1. hour 2. day 3. week 4. month 5. year
How many bags per kiln? |__|__|__|$_{(25.4)}$
26. How many or which months are the dry season here? _____(26.0)
27. How large of an area do you cut for one kiln (show length of one side and make approximation) |__|__|__|__|__| m^2 $_{(27.0)}$
28. How long does it take you to cut and stack the trees for one kiln?
              |__|__|__| |__|(TIME)$_{(28.0)(28.1)}$
    TIME 1. hour 2. day 3. week 4. month 5. year
How many hours per day do you spend on this production during this time?
from _____ to _____ (times) or |__|__| hours per day $_{(28.2)}$
How long does it take you to cover one kiln?
              |__|__|__| |__|(TIME)$_{(28.3)(28.4)}$
How long does it take you to burn one kiln?
              |__|__|__| |__|(TIME)$_{(28.5)(28.6)}$

How many hours per day do you spend on this production during this time?
from _____ to _____ (times) or |__|__| hours per day (28.7)
29. Do you have anyone helping you with this work? (0. no 1.yes) |__|(29.0)
   If yes, how many of each type:
TYPE	FAMILY	NON-FAMILY	
cutting trees	|__|__|	|__|__|	(29.1)(29.2)
stacking wood	|__|__|	|__|__|	(29.3)(29.4)
covering kiln	|__|__|	|__|__|	(29.5)(29.6)
burning mkaa	|__|__|	|__|__|	(29.7)(29.8)

   How much money do you spend on Labour for an average kiln?
            Tsh |__|__|__|__|__|(29.9)
30. How much do you sell your mkaa for?   UNIT
   DRY SEASON Tsh |__|__|__|__|__| per |__|(30.0)(30.1)
   WET SEASON Tsh |__|__|__|__|__| per |__|(30.2)(30.3)
   Last year, dry season Tsh |__|__|__|__|__| per |__|(30.4)(30.5)
UNIT: 1.large bag 2. small bag 9. other (specify)_____(30.6)
31. Are gunia provided by the buyer? (0. no 1.yes) |__|(31.0)
32. Who fills the gunia? (1. you 2. buyer/employees of buyer) |__|(32.0)
==

PART 3 – KUNI

33. Do you cut or collect kuni for sale on a regular basis? (0. no 1.yes)
    |__|(33.0)
    IF NO ANSWER ONLY QUESTIONS 36, 37, and 40
34. How long have you been cutting/collecting kuni for sale?
          |__|__|.|__|__|years OR since |__|__|__|__|(year)(34.0)
35. Do you cut/collect more kuni for sale in the dry season or the wet
    season?   (1. dry 2. wet) |__|(35.0) Why?_____(35.1)
36. How much do you sell?
                  UNIT   TIME
   DRY SEASON |__|__|__| |__| per |__|(36.0)(36.1)
   WET SEASON |__|__|__| |__| per|__|(36.2)(36.3)
       TIME 2. day 3. week 4. month 5. year
       UNIT: 3. Log 4. large piece 5. small piece 6. Bundle 7. Kg 8. Tonne
       9. other (specify)_____(36.4)
Weight of one unit (1) |__|__|__|__|kg (2) |__|__|__|__|kg
(3) |__|__|__|__|kg (4) |__|__|__|__|kg (5) |__|__|__|__|kg
AVG |__|__|__|__|kg (36.5)

37. How many or which months are the dry season here?_____ (37.0)
38. In the dry season, how much time do you spend on cutting or collecting wood? from _____ to _____ (times) or |__|__| hours per day (38.0)
AND |__| days per week (38.1)
In the wet season, how much time do you spend on cutting or collecting wood? from _____ to _____ (times) or |__|__| hours per day (38.2)
AND |__| days per week (38.3)
39. Do you have anyone helping you with this work? (0. no 1.yes)|__|(39.0)
   If yes, how many of each type:

TYPE	FAMILY	NON-FAMILY
cutting trees	|__|__|	|__|__|(39.1)
stacking wood	|__|__|	|__|__|(39.2)
splitting logs	|__|__|	|__|__|(39.3)

How much money do you spend on Labour for your kuni business in an average month   Tsh |__|__|__|__|__|__|(39.4)

40. How much do you sell your kuni for?
                                             UNIT
       DRY SEASON Tsh |__|__|__|__|__| per |__|(40.0)
       WET SEASON Tsh |__|__|__|__|__| per |__|(40.1)
   Last year, dry season Tsh |__|__|__|__|__| per |__|(40.2)
   UNIT: 3. Log 4. large piece 5. small piece 6. Bundle 7. Kg 8. Tonne
   9. other (specify)_____(40.3)